Illustrated Dictionary of Practical Astronomy

Springer
London
Berlin
Heidelberg
New York
Barcelona
Hong Kong
Milan
Paris
Singapore
Tokyo

Chris Kitchin

Illustrated Dictionary of Practical Astronomy

With 273 Figures, including 3 in colour

Springer

British Library Cataloguing in Publication Data
Kitchin, Christopher R. (Christopher Robert), 1947–
Illustrated dictionary of practical astronomy
1. Astronomy – Dictionaries 2. Astronomy – Handbooks, manuals, etc.
I Title
522′.03
ISBN 1852335599

Library of Congress Cataloging-in-Publication Data
A catalog record for this book is available from the Library of Congress

ISBN 1-85233-559-9 Springer-Verlag London Berlin Heidelberg
a member of BertelsmannSpringer Science+Business Media GmbH
http://www.springer.co.uk

Printed in Singapore

Typeset by EXPO Holdings, Malaysia
Printed and bound by Kyodo Printing Co. (S'pore) Pte. Ltd., Singapore
58/3830-543210 Printed on acid-free paper SPIN 10844367

For Jess and Spruce

Preface

The purpose of this book is two-fold. Firstly to provide a handy quick source of reference of the terms, techniques, instruments, formulas, processes, etc., for practising observers, whether it is their first look through someone else's small telescope, or whether they have spent decades building their own instruments, observing with them and are regularly producing results to rival those of the professionals. It is not principally aimed at professional observers, but in order to be sufficiently comprehensive for its intended audience, many entries have been included which will be relevant at that level. In particular though, the more esoteric parts of spectroscopy have not been included. References to specific observatories are included if their instrumentation includes optical telescopes over 1 m in diameter or radio dishes over 20 m. Individual entries for telescopes of 4 m or over are included, and for the larger radio instruments, plus other telescopes that may be of interest for historical or other reasons (for example the 1m Yerkes refractor). Spacecraft have generally not been included (apart from the Hubble space telescope) because their short working lives mean that most current spacecraft will no longer be operating by the time that this book is published. Also the names of spacecraft are frequently changed after launch making it difficult to identify which mission is which. References to commercial organisations, and to some widely available commercial products have been included, but an intending purchaser should always obtain up-to-date information.

The second purpose of this book is to give sufficient information for many of the techniques, etc. referred to within it, to be understood and even used. For example, rather than just give a simple definition of precession of the equinoxes, as might a normal dictionary, the entry includes the formulae to enable the observer to calculate the effects of precession. Similarly, web page addresses are given for many of the organizations referenced in the book, so that the interested reader can quickly obtain more information, and so on. There is obviously a limit to the amount of information that can be encompassed in a dictionary/handbook format, so more complex topics will require further reading, and a bibliography is included in the Appendix.

Cross references are given in italics within the entries, and additional cross references are given at the end of many entries.

I hope that you find this book useful, and that when you look up a topic, the entry gives you the information that you need. However, I would be happy to hear of extra material that might be included in any future edition, whether it be additions or changes to existing entries, or complete new entries.

I wish you all clear skies and happy observing!

C.R. Kitchin
Hertford, April 2002

Acknowledgements

The author and publishers would like to thank the following for the use of images reproduced from other Springer titles:

Mel Bartels for the upper-right image on page 203 which was taken from *Amateur Telescope Making* (edited by Stephen F. Tonkin).

Denis Buczynski for the lower image on page 67 which was taken from *The Modern Amateur Astronomer* (edited by Patrick Moore).

Steven R. Coe for the images on pages 68 (upper right & lower left), 86 (lower left), and 163 (middle left & lower left) which were taken from his book *Deep-Sky Observing*.

Robert W. Forrest for the image on page 201 which was taken from *The Modern Amateur Astronomer* (edited by Patrick Moore).

Alan W. Heath for the upper-left image on page 169 which was taken from *Small Astronomical Observatories* (edited by Patrick Moore).

Steven Lee for the images on pages 23 and 203 (upper left) which were taken from *Amateur Telescope Making* (edited by Stephen F. Tonkin).

Martin Mobberley for the images on pages 33 (middle & lower), 81 (lower), 86 (upper left), and 169 (lower left & lower right) which were taken from his book *Astronomical Equipment for Amateurs*.

Rod Mollise for the images on pages 60 (upper right & lower) and 114 which were taken from his book *Choosing and Using a Schmidt-Cassegrain Telescope*.

Patrick Moore for the images on pages 3 (upper left), 5, 28, 112, 120 (upper), 133 (upper left), 148, 154 (lower), 155, 163 (upper right), 204 (lower left & lower right), 239 (upper), and 255 (lower) which were taken from his book *Eyes on the Universe*.

David Ratledge for the middle & lower images on page 38 which were taken from his book *The Art and Science of CCD Astronomy*.

Chuck Shaw for the image on page 188 which was taken from *Amateur Telescope Making* (edited by Stephen F. Tonkin).

Gil Stacy for the lower-right image on page 68 which was taken from *Amateur Telescope Making* (edited by Stephen F. Tonkin).

Credits for other images are shown in their individual captions.

A

AAS
See *American Astronomical Society.*

AAT
See *Anglo Australian telescope.*

AAVSO
See *American Association of Variable Star Observers.*

Abastumani astrophysical observatory
An optical observatory on Mount Kanobili, Republic of Georgia, altitude 1700 m.

Abell catalog
A catalog of clusters of galaxies.

aberration
1 Optical aberration; one of six faults occurring in images. For details see the individual entries: *astigmatism; chromatic aberration; coma; distortion; field curvature* and *spherical aberration;* See also: *achromatic lens; apochromatic lens; anastigmatic optics; aplanatic optics; parabolic mirror; Schmidt camera.*
2 Stellar aberration; a small change in the positions of objects in the sky arising from the Earth's motion through space. Objects are moved by up to 20″ towards the instantaneous direction in space of the Earth's motion. Over a year the object therefore traces out a small ellipse in the sky. The effect of aberration needs to be corrected when determining positions to high levels of accuracy. See also: *parallax; diurnal aberration; elliptical aberration.*

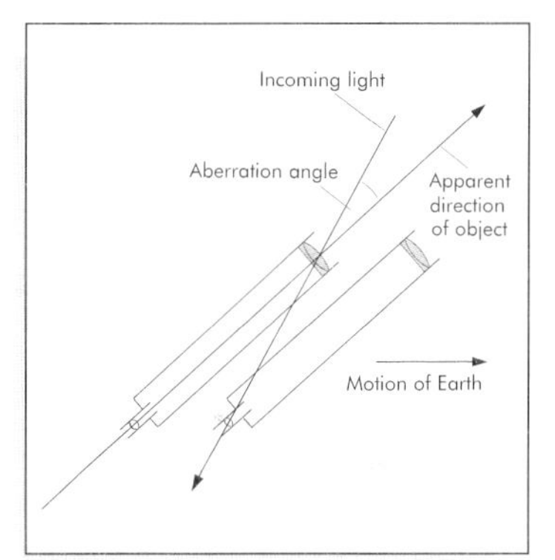

Stellar aberration.

absolute magnitude
A measure of an object's intrinsic brightness, usually symbolized by, M. It is defined as the *apparent magnitude* if the object were 10 pc away. It is related to apparent magnitude, m and distance, D, (in pc) by $M = m + 5 - 5\log_{10} D$.
Some example values are:

	M
Supernova type Ia	−19
α Orionis (Betelgeuse)	−6
Sirius A	+1.4
Sun	+4.6
Faintest white dwarfs	+16
Jupiter (at its brightest)	+26

absorption, atmospheric
See *atmospheric extinction.*

absorption line
A dark line appearing in the *spectrum* of an object. The line arises from the absorption of photons at a specific *wavelength* by ions, atoms or molecules. See also: *emission line.*

A

achromatic doublet
See *achromatic lens.*

achromatic lens
A lens in which the *chromatic aberration* has been reduced compared with that to be expected from a simple lens. The first achromat was invented by Chester Moor Hall in 1733 and manufactured by John Dolland in 1754 and greatly enhanced the usefulness of *refractors*. Achromats have two or more lenses formed from different types of glass, the chromatic aberration of one lens being made equal and opposite to that of the other, thus largely cancelling out its effects for the combination. See also: *aberration (optical); Fraunhofer lens; Littrow condition; Clairault condition; crown glass; flint glass; secondary spectrum; apochromat; orthoscopic.*

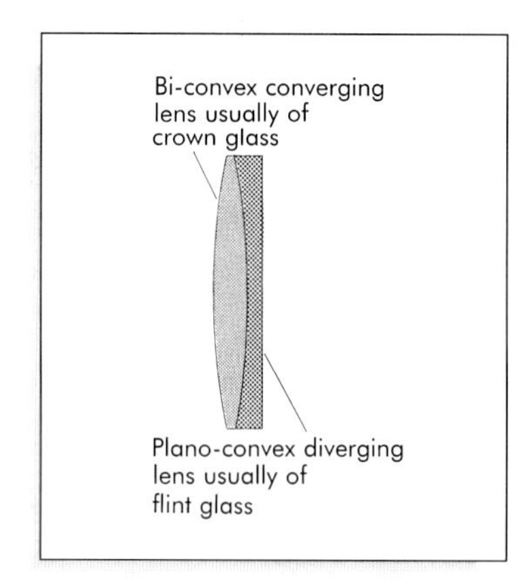

Achromatic lens.

achromatic Ramsden eyepiece
See *Kellner eyepiece.*

active atmospheric compensation
See *adaptive optics.*

active optics
See panel (page 3).

active supports
A support system for an optical component, usually a large mirror, wherein the individual supports can be moved to vary the force applied to the mirror. The supports are changed under computer control in order to apply stresses to the mirror that counteract the stresses arising from the changing gravitational loading on the mirror as the telescope moves around the sky. The mirror thereby maintains its correct shape. Active supports with response times in milliseconds are also required for *adaptive optics;* See also: *active optics.*

adaptive optics
See panel (page 4).

ADC
See *Astronomical Data Center.*

aeon
A unit of time equal to 10^9 years.

aerial
See *antenna.*

active optics

A large thin mirror or a large mirror made from smaller segments, whose shape is kept to the required accuracy by frequent adjustment of its supports. The mirrors of small telescopes may be made sufficiently rigid that the distortion induced by gravitational stresses (or due to wind in the case of radio telescopes) is insignificant. But this is not practical for large mirrors. The shape of a large mirror is therefore monitored by observing a guide star, or by laser interferometry. The computer-controled supports of the mirror are then adjusted at intervals of a second or so to maintain its shape to the required accuracy. See also: *active supports.*

The Keck I telescope with the hexagonal segments of its primary mirror clearly seen.

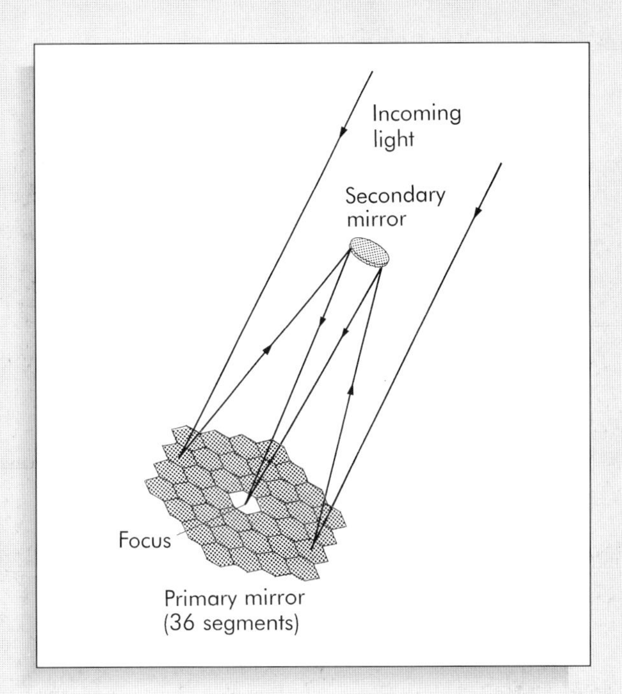

The optical layout of the Keck telescope's primary mirrors

One of the primary mirrors of the European Southern Observatory's Very Large Telescope (VLT) the mirror is thin compared with its 8.2 m diameter and so an active optics system is required to maintain its correct shape. (Image reproduced by courtesy of the European Southern Observatory).

One of the primary mirror cells of the European Southern Observatory's Very Large Telescope (VLT) showing the numerous supports of the active optics system. (Image reproduced by courtesy of the European Southern Observatory).

adaptive optics

An optical system that enables a large Earth-based telescope to work at or close to its *Rayleigh resolution*. The adaptive optics system compensates for the distortions induced into the image by the Earth's atmosphere by producing equal and opposite distortions. It must respond in a few milliseconds to the changing effects of the atmosphere. A guide star is required near to the object whose image is to be corrected. If such a star is not available, some systems produce an artificial star by shining a powerful laser upwards to cause sodium atoms to glow at a height of about 90 km. The atmospheric distortions induced in the image of the guide star are then monitored. A small thin or segmented mirror on *active supports* is placed in the light beam from the telescope and its shape distorted to compensate for the distortions produced by the atmosphere. See also: *tip–tilt mirror; isoplanatic patch; shearing interferometer; Hartmann sensor.*

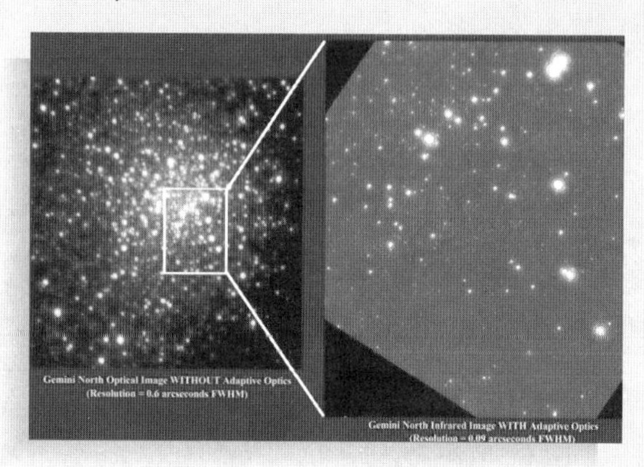

Adaptive optics at work. Images of NGC 6934 from the Gemini North telescope, on the left without adaptive optics, and on the right with adaptive optics. (Images courtesy of Gemini Observatory and the University of Hawaii Adaptive Optics Group).

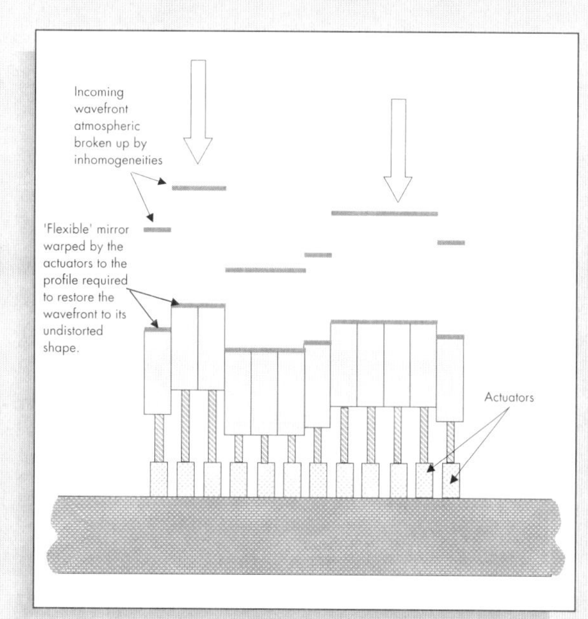

The 'flexible' mirror required to restore the wavefront to its undistorted shape. The mirror may be made of segments as illustrated here, or be a single thin mirror that bends under the forces applied by the actuators.

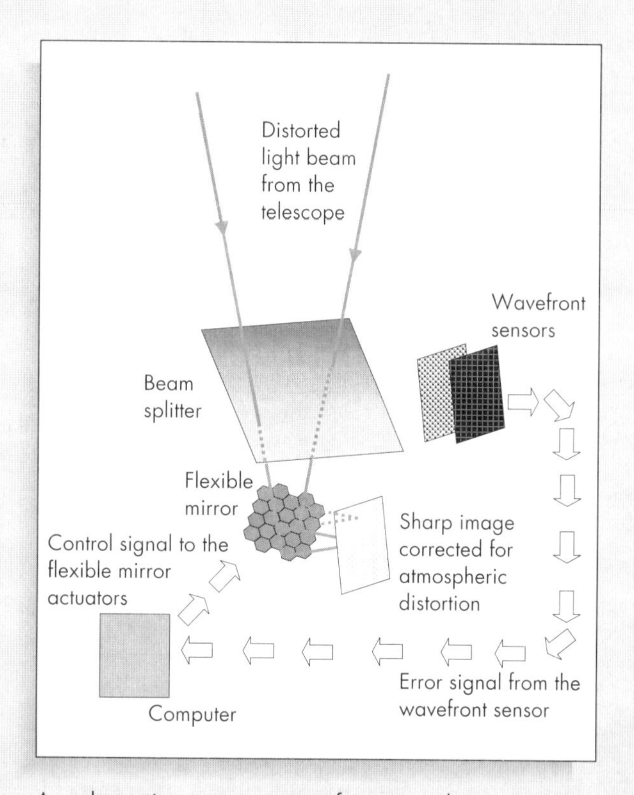

A schematic arrangement for an adaptive optics system. The beam splitter sends a small (10%) proportion of the distorted light beam from the main telescope to a wavefront sensor. The sensor detects the distortions in the wavefront and generates an error signal. The error signal is then used by the computer to control the actuators for the flexible mirror to warp the mirror in such a way that the wavefront is restored to its correct shape after reflection.

aerial telescope

A design of *refractor* used in the mid-seventeenth century. To reduce the effects of *aberrations* from a simple lens, immense *focal ratios* were used, and the telescopes were up to 50 m long. The *objective* was mounted on a mast and raised and lowered by ropes. The observer held the *eyepiece*, which was connected to the objective by a thin line, and kept the line taut in order to align the objective and eyepiece.

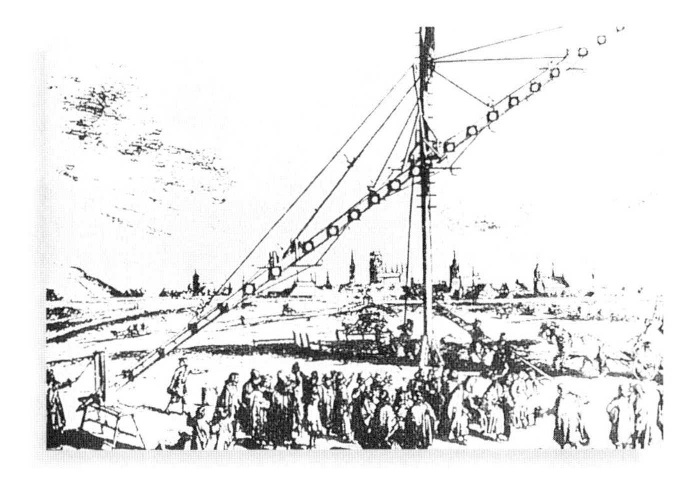

An aerial telescope.

afocal imaging

The use of a conventional *digital* or *single lens reflex camera* (i.e. with its normal lens in place) to obtain an image through a normal telescope (i.e. one with an eyepiece in place). The camera and telescope are both set to infinity, and the camera held to look through the eyepiece. It gives a magnified image compared with direct imaging (i.e. with the telescope *objective* replacing the camera lens) or compared with *eyepiece projection*.

afocal optics

An optical system in which a parallel light beam entering the system, produces a parallel exit beam. This is the normal setting for a telescope used visually since it is the eye which actually produces the image. See also: *confocal*.

aircraft trail

See *trails*.

air glow

A component of the *sky background* that is due to emission from the atoms and molecules in the Earth's atmosphere. See also: *aurora*.

air mass

A synonym for *air path*.

air path

The length of the Earth's atmosphere traversed by a light ray to reach the observer. *Atmospheric extinction* is proportional to the length of the air path. Taking a value of unity for an object at the zenith and an observer at sea level, the air path is given approximately by:

$$\text{airpath} \approx \frac{1}{\cos z}$$

for zenith distances (z) up to about 60°. The air path is reduced by about 12% for each 1,000 m of altitude above sea level.

A

Airy disk

The central region of the interference pattern resulting from *diffraction* at the edges of a telescope's *objective*. Because of diffraction, the image of a point source, such as a star, even in an aberration-free telescope, without atmospheric distortion, is not a point, but is a small circle with blurred edges, surrounded by faint rings. The central circle is the Airy disk. Its radius (to the center of the first dark fringe) is $1.22\lambda/D$ *radians* (where λ is the *wavelength* and D the objective diameter). For optical telescopes, this equates to 0.12/D seconds of arc (where D is in meters), and this is called the *Rayleigh resolution* of the telescope. For a dish-type radio telescope the Airy disk is the same as the main lobe of the *polar diagram*, and the fringes are the *side lobes*.

The Airy disk and surrounding fringes.

Algonquin telescope

A 46m steerable radio dish at Algonquin in Canada.

aliasing

A false signal obtained when a rapidly varying quantity is sampled at too slow a rate. A spurious low frequency component to the observed signal results from beating between the sampling rate and the higher frequencies in the original signal. Thus a pulsar with a period of 100 ms, sampled at 167 ms intervals, would appear to have a period of 500 ms. See also: *Nyquist frequency*.

alidade

See *astrolabe*.

alignment

The setting up of a telescope mounting so that objects may be *tracked* correctly.

Modern small *alt-azimuth* mountings normally come with appropriate software to align the telescope, and this normally involves finding two bright stars with known positions and following the manufacturer's instructions.

An *equatorial mounting* needs to be aligned so that its *polar axis* is precisely parallel with the Earth's rotational axis. If the telescope has a *cross-wire eyepiece* and *setting circles*, the initial setting-up can be done by setting the telescope to the coordinates of the pole star, then finding the pole star and centring it on the cross wires by moving the polar axis in azimuth and altitude. Even without setting circles, aligning the telescope by eye to be parallel with the polar axis and then finding Polaris will set the mounting up to within a degree or two. Some commercially produced telescopes are provided with a rifle sighting scope which may be attached to the mounting in order to sight on the pole star in a similar way. An arrangement of this sort is almost essential for a portable telescope if considerable amounts of time are not to be wasted every time it is used.

A permanently installed telescope mounting can be aligned much more precisely after this initial setting up, using observational tests. Corrections to the altitude of the polar axis are found by observing a star of *declination* between 30° and 60°, about 6 hours east or west of the observer's meridian. The star should be centered on the cross wires of a high power eyepiece, and tracked using the telescope drive. If, after a while, a star to the east of the meridian is observed to drift northwards in the eyepiece, then the altitude of the polar axis is too high. A star to the west of the meridian would drift to the south in the same circumstances. If the easterly star drifts south, or the westerly star, north, then the polar axis is set to too low an altitude. After a suitable adjustment to the axis, the procedure is repeated, until any remaining drift is acceptable. Alignment will clearly have to be done much more precisely if the observer intends, say, to undertake photography with exposures of several hours rather than visual work. A similar procedure may be used to align the polar axis in azimuth. Two stars are selected which are ten or twenty degrees north and south of the equator, and differing in *right ascension* by a few minutes of arc. When the stars are within an hour or so of transiting the observer's meridian, with the drive OFF, the telescope is set just ahead of the leading star (the one with the smaller right ascension). A stopwatch is started as that star transits the vertical cross wire. The telescope is then moved to the declination of the second star, and the time interval until that star transits is measured. The correct time interval between the two transits can be found from the difference in their right ascensions. If the measured time interval is too small, then the polar axis is aligned to the east, if the time interval is too large, then the polar axis is to the west of its true direction. Adjustments to the position of the polar axis are made and the procedure again repeated until satisfactory.

allowed spectrum line

A 'normal' *spectrum line*, i.e. one that arises without breaking any of the rules governing *transitions*. See also: *forbidden line; intercombination line.*

ALMA

See *Atacama Large Millimeter Array.*

almanac

A listing, usually produced annually, of the changing properties of astronomical objects (such as the positions of the Sun, Moon and planets, phases of the Moon, positions of Jupiter's satellites, eclipses, *sidereal* and *solar times*, etc.). The best known is the *Astronomical Almanac* published jointly by HMSO and the US government. See also: *ephemeris.*

The *Astronomical Almanac.*

almucantar

A device for measuring the *altitude* and *azimuth* of an object. See also: *astrolabe.*

alt–az

Abbreviation of *altitude* and *azimuth*. Usually used to refer to the *alt–az telescope mounting.*

alt–az telescope mounting

A mounting for a telescope where the telescope moves in *altitude* and *azimuth*. Most large telescopes (*optical* and *radio*) and increasing numbers of smaller telescopes now have mountings of this type, even though computer control of the drive motors is needed in order to set the telescope onto an object and then to *track* it across the sky. See also: *Dobsonian mounting; equatorial mounting; wedge.*

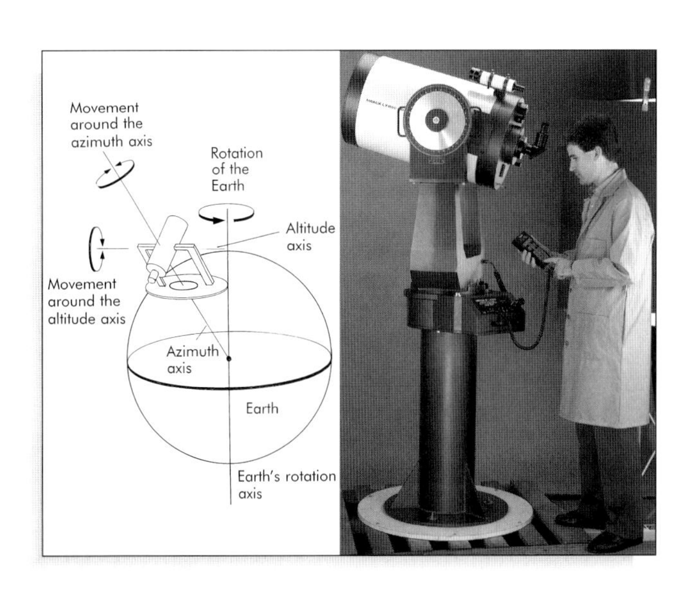

The alt–azimuth telescope mounting showing its alignment with the Earth (left); and an actual mounting for a 16-inch telescope (right). (Right image reproduced by courtesy of Meade Instrument Corp.)

altitude

1 The angle of an object in the sky above the horizon, measured from 0° to 90°, symbol, a. It is related to the *RA* (α), *Dec* (δ), of the object, the observer's latitude (ϕ) and the local *sidereal time* (LST) by $\sin a = \sin \delta \sin \phi + \cos \delta \cos (\text{LST} - \alpha) \cos \phi$. NB LST and RA must be converted to degrees before inserting into this formula. See also: *azimuth; zenith distance.*

2 The height above sea level on the Earth, or above the reference data surface on other planets.

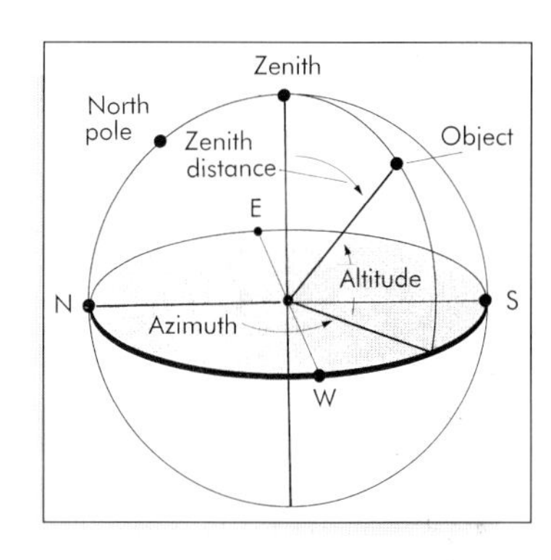

The definition of altitude.

aluminizing

The application of a thin reflecting coat of aluminum to the surface of an optical component to produce a mirror. The substrate is usually glass or a low-expansion material such as quartz, *Zerodur*, or *ULE* whose surface has been *figured* to the required shape. The substrate is placed into a vacuum chamber and coils of thin aluminum wire heated above the surface until they evaporate and deposit onto the substrate. An *over coating* of silicon dioxide is then often added for protection. Other metals may be used, especially for use in the ultraviolet. The reflectivity of aluminum will decline from over 90% to 75% or so in a year, and the mirror then has to be cleaned of its previous layer of aluminum and re-coated. The aluminum coat can be cleaned in between re-coatings by (for example) drifting carbon dioxide snow across it. See also: *speculum metal; silvering; sputtering; mirror coating; cleaning.*

AMANDA neutrino detector

A *neutrino detector*, based upon detecting the Čerenkov radiation from high-energy electrons produced in ice within the Antarctic icecap. The name is an acronym from Antarctic Muon And Neutrino Detector Array.

American Association of Variable Star Observers

A USA-based but international association for observers interested in variable stars, headquarters at Cambridge, Massachusetts. See http://www.aavso.org/.

American Astronomical Society

A society for amateur and professional astronomers based in Washington, DC. See http://www.aas.org/.

amplitude

The difference between maximum and minimum of a varying signal. Usually used in the context of variable stars. For example the amplitude of RR Lyrae stars lies between 0.2^{m} and 2^{m}.

analemma

See *equation of time.*

anastigmatic optics

An optical system in which a flat focal surface is retained and *astigmatism* is corrected for at least two distances from the *optical axis.* See also: *aberration (optical); field curvature; orthoscopic.*

angle

An angle is defined (in units of *radians*) as; the length of the curved edge of the sector of a circle enclosed by the angle divided by the radius of the circle. A complete circle thus has an angle of 2π radians, Since this is also 360°, we have 1 rad = 57° 17' 45". See also: *degree; minute of arc; second of arc; hour; minute; second; solid angle.*

angle of dip

See *dip of the horizon.*

Anglo Australian observatory

An optical observatory (altitude 1200 m) at Siding Springs in Australia. It houses the 3.9 m *Anglo Australian telescope* and the 1.2 m *UK Schmidt camera.*

Anglo Australian telescope

A 3.9 m *Ritchey-Chrétien* telescope at the *Anglo Australian observatory.* For images see http://www.aao.gov.au/images.html/.

angstrom

A unit of length equal to 10^{-10} m, symbol, Å. Named for the Swedish astronomer Anders Ångström (1814–74). The preferred SI unit is the *nanometer* (10^{-9} m, symbol, nm, 1 nm = 10 Å). However, the unit is still in widespread use within astronomy especially as a unit for the *wavelength* of light. See also: *micron.*

angular diameter

The diameter of an object as seen in the sky and measured in angular units (*radians; degrees; arc minutes; arc seconds; hours; minutes; seconds*). The angular diameter must be combined with the distance of the object to determine its true diameter. Objects at different distances (e.g. the Sun and the Moon) can have similar angular diameters, but very different actual sizes.

angular estimate

A useful guide to angular size for naked eye work is that the clenched fist held at arms length has a width of about 8° to 9°, likewise the width of the index finger is about 2°.

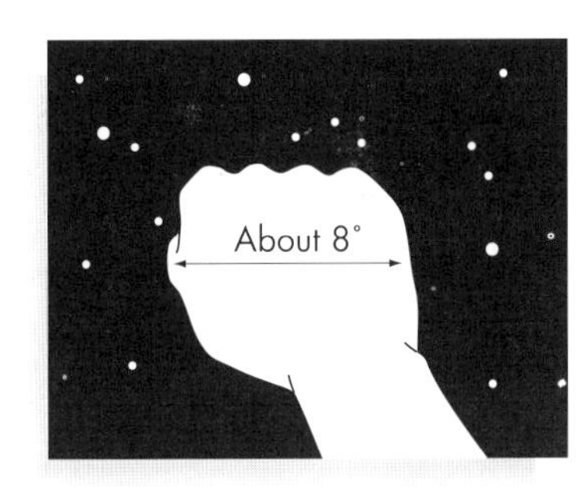

Estimating angles in the sky.

angular resolution

See *Rayleigh resolution* and *interferometer resolution.* See also: *Airy disk.*

annealing

A part of the manufacturing process when glass or similar materials are being cast to form the blanks for mirrors or lenses. The blank has to be cooled down very slowly in order to allow stresses to dissipate. These stresses might otherwise deform the blank when cold or even cause it to shatter. For large mirror blanks, the annealing process may last for several months.

anomalous Zeeman effect

See *Zeeman effect.*

antenna

The component of a *radio receiver* or transmitter that is used to detect or transmit the radio signal. It is also known as an *aerial.* For radio telescopes it is often a *half-wave dipole*, which comprises two conducting strips each a quarter of the wavelength long. The radio signal induces an electric current into the half-wave dipole that is then amplified and converted to the desired form by the remainder of the radio receiver. The half-wave dipole is called the active element of the antenna. Additional parasitic elements may be added to reflect or focus the radio signal onto the active element. These consist of conducting strips that are not electrically connected to the active element and placed before the half wave dipole (to focus) or after (to reflect) the signal onto it. This combination of active and parasitic elements results in the familiar *Yagi* or *parasitic antenna* used to receive domestic TV. Such antennas are also often used at the focus of dish-type radio telescopes to detect the signal.

antenna gain
See *gain*.

antenna pattern
See *polar diagram*.

antenna tapering
The use of a *feed* for a *radio telescope* that intercepts only the *main lobe* of the *beam pattern*. It reduces the overall sensitivity and resolution of the instrument, but that is likely to be worth the trade off against the reduction in the confusion cause by *side lobes*. It is a similar technique to *apodization* at optical wavelengths.

antenna temperature
The temperature of a resistor that produces the same level of thermal noise as the signal from the antenna. It is a measure of the strength of the incoming signal, not the actual temperature of the antenna.

anti-blooming
The use of extra electrodes, known as drains, within a *CCD* that are able to bleed off electrons from *pixels* approaching *saturation*. See also: *blooming*.

anti-reflection coating
See *blooming*.

Antoniadi scale
A scale used to quantify the *seeing* conditions during observations. It was first devised by Eugène Antoniadi (1870–1944). There are five levels of seeing:
I Perfect
II Some turbulence but perfect for several seconds at a time
III Moderate, deteriorating to poor at intervals
IV Poor, with constant turbulence
V Appalling, observations almost impossible.

Antu
See *very large telescope*.

Apache point observatory
An observatory housing 3.5m, 2.5m and other large optical telescopes in the Sacramento mountains, New Mexico, altitude 2800 m. It is undertaking the *Sloan digital sky survey*.

aperture
The diameter of the clear part of the *objective* of a telescope. This is usually the same as the diameter of the objective, although in some cheap telescopes internal stops will reduce the effective aperture below that of the objective. Also used in a similar fashion for eyepieces and camera lenses.

aperture efficiency

The efficiency of a *radio telescope* in receiving radiation. For dish-type telescope it is the ratio of the effective area to the actual area of the dish. Typically it will have a value in the region of 60% to 70%. See also: *beam efficiency.*

aperture ratio

See *f-ratio.*

aperture synthesis

See panel (pages 12–13).

aplanatic optics

An optical system in which both *spherical aberration* and *coma* have been corrected. Many modern large optical telescopes use the aplanatic *Ritchey-Chrétien* design. See also: *aberration (optical); orthoscopic.*

apochromat

A color-corrected lens with three or more components and a higher level of correction of *chromatic aberration* than the standard *achromat.* In the achromat two different wavelengths have the same focus position. In the apochromat three different wavelengths have the same focus position, and in the *super-apochromat* four wavelengths have the same focus. See also: *aberration (optical); orthoscopic.*

apodization

A process whereby the normal performance of an instrument is degraded in one respect in order to enhance its performance in another respect. For example a neutral density filter, which varies in a Gaussian fashion from transparent at the center to opaque at the edges, when placed over the objective of a telescope, will double the size of the *Airy disk*, but suppress the surrounding fringes. Despite the reduction in the nominal resolution arising from this, the suppression of the fringes will lead to significantly more detail being seen in low contrast extended objects like planets and nebulae. See also: *antenna tapering.*

Apogee Instruments

A commercial firm manufacturing CCD cameras.

apparent field of view

The *field of view* of an *eyepiece.* See also: *true field of view.*

apparent magnitude

A measure of the brightness of an object as it appears in the sky, symbolized by m. The roots of the system go back to Hipparchus' star catalog of 150 BC. That catalog characterized stars as class 1 (brightest) through to class 6 (those just visible). The modern system is due to Norman Pogson (1829–91) and has magnitude 6 stars as those just visible to the eye. Each magnitude difference then corresponds to a change in brightness by a factor of ×2.512 ($= \sqrt[5]{100}$). Magnitude 5 stars are thus 2.512 times brighter than magnitude 6. Magnitude 4 stars are 6.3 ($= 2.512^2$) times brighter than magnitude 6, magnitude 3 stars are 15.9 ($= 2.512^3$) times brighter and so on. The brightest objects require negative magnitudes to fit onto the scale (e.g. Sirius, m = –1.45, full Moon, m = –12.7). Objects fainter than magnitude 6 have magnitudes of 7 or more (e.g. the star at the center of the Ring Nebula, M57, is of magnitude 15). The relationship between the brightnesses of objects and their apparent magnitudes is given by Pogson's equation:

$$m_1 - m_2 = -2.5\log_{10}\left(\frac{B_1}{B_2}\right),$$

aperture synthesis

A method of observing using an *interferometer* whereby the equivalent of observing an object with a conventional telescope with a diameter equal to the maximum separation of the components of the interferometer is synthesized. Also called Earth rotation synthesis. The technique was first applied at radio wavelengths, but is now starting to be used in the infrared and visible regions. It relies upon the Earth's rotation to change the orientation in space of a pair of components of an interferometer through 360° over a 24-hour period. The two components thus trace out an annulus with a diameter equal to their separation and a thickness equal to their *apertures* over that period. If one component is then moved through its own diameter, the next annulus can be synthesized in the next 24-hour interval. Over a period of time, by moving the two components from being next to each other to their maximum separation, adjacent annuli can be synthesized to give the effect of observing with a normal telescope with an aperture equal to the maximum separation. The conversion of the output of an aperture synthesis system to an image of the object is accomplished through the use of *Fourier transforms.*

The technique can only be used on objects that will remain unchanged over the period of observation. The process may, however, be speeded-up by using many components. For example an interferometer with 5 components could provide 10 different separations simultaneously, one with 10 components would provide 45 separations, etc. Also, in practice, only 12 hours of observation are needed since the other 12 hours can be reconstructed within the computer recording the observations. Even shorter periods of observation can be achieved by using several arms to the interferometer, oriented at different angles to each other, Thus the *very large array* (VLA) for example has a Y-shaped configuration and twenty seven 25m diameter radio dishes with a maximum separation of 36 km.

If all the annuli required for the equivalent normal dish are synthesized, then it is a filled-aperture system. If some annuli are omitted, then it is an un-filled aperture system. This latter is common for the larger base-line systems such as *MERLIN*, which has a maximum separation of 217 km. Unfilled aperture systems synthesize the resolution but not the sensitivity of the equivalent normal telescope. See also: *very long base-line interferometry; synthetic aperture radar; uv plane.*

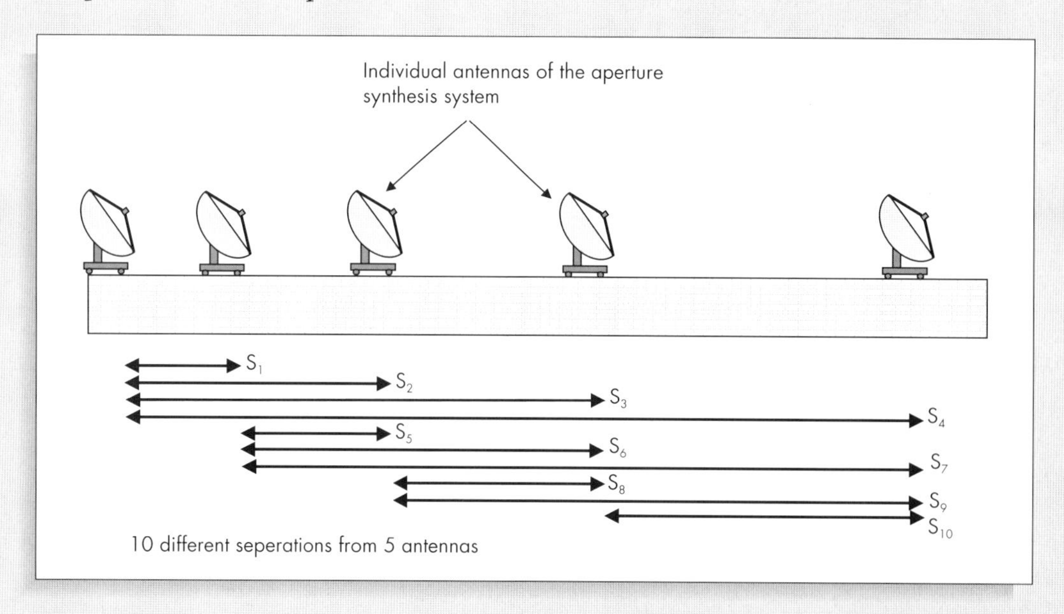

The use of several antennas in an aperture synthesis system allows many separations to be observed simultaneously

The Very Large Array (VLA) radio aperture synthesis system looking Southwest at twilight (Image reproduced by courtesy of Dave Finley, NRAO/AUI).

The Dominion Radio Astrophysical Observatory's synthesis telescope at Penticton, British Columbia, operated by the National Research Council Canada (Image reproduced by courtesy of National Research Council Canada).

where m_1 and m_2 are the magnitudes of the two objects, and B_1 and B_2 are their brightnesses or energies. The scale ranges from the Sun ($m = -26.7$) to the faintest objects currently detectable ($m = +28$), and this corresponds to a difference in brightnesses by a factor of $\times 10^{22}$. See also: *absolute magnitude; limiting magnitude; combined magnitude.*

apparent sidereal time

The *hour angle* of the actual *First Point of Aries*. See also: *sidereal time; mean sidereal time; equation of the equinoxes.*

apparition

The time period over which a moving object such as a planet or comet is visible from Earth.

appulse

The close passage in the sky of a moving object such as a planet to another prominent moving or stationary object in the sky. See also: *conjunction; eclipse; transit; occultation.*

arc minute

A measure of angle equal to 1/60 of a *degree*, symbol, ′. See also: *arc second; radian.*

arc second

A measure of angle equal to 1/3600 of a *degree*, symbol, ″. See also: *arc minute; radian.*

Arecibo radio telescope

The largest dish-type *radio telescope* in the world, located at Arecibo in Puerto Rico. It is 305 m in diameter and is stationary. The rotation of the Earth, however, enables it to scan objects crossing its zenith, and objects up to 20° away from the zenith may be observed and tracked by moving the *feed* around the focal plane of the reflector.

Argelander step method

A variant of the *fractional estimate* method of obtaining visual apparent magnitudes. See also: *Pogson step method.*

argument of perihelion

A synonym for longitude of perihelion (see *orbital elements*).

array

A collection of simple radio *antennas* such as *half wave dipoles* or *Yagis* whose outputs are combined to give greater sensitivity and resolution. The two common arrangements are the collinear array where half wave dipoles are strung out in a line, with separations of $\lambda/2$ and with the dipoles are aligned along the array axis, and the broadside array that has a similar arrangement except that the dipoles are perpendicular to the array axis. The term may also be used for assemblies of several dish-type antennas.

artificial guide star

An imitation star produced by shining a laser upwards to cause sodium atoms at a height of around 90 kilometers in the Earth's atmosphere to glow, for use in *adaptive optics.*

ascending node

See *node.*

Ash domes

A commercial firm manufacturing domes for telescopes.

Asiago observatory

An optical observatory in the Asiago highlands, Italy, at an altitude of 1000 m.

ASP

See *Astronomical Society of the Pacific.*

aspect

The geometrical position of a planet, etc. with respect to the Earth and Sun. (Tabulated annually in the Astronomical *Almanac.*) See also: *inferior and superior conjunction; opposition; greatest elongation; quadrature; phase angle.*

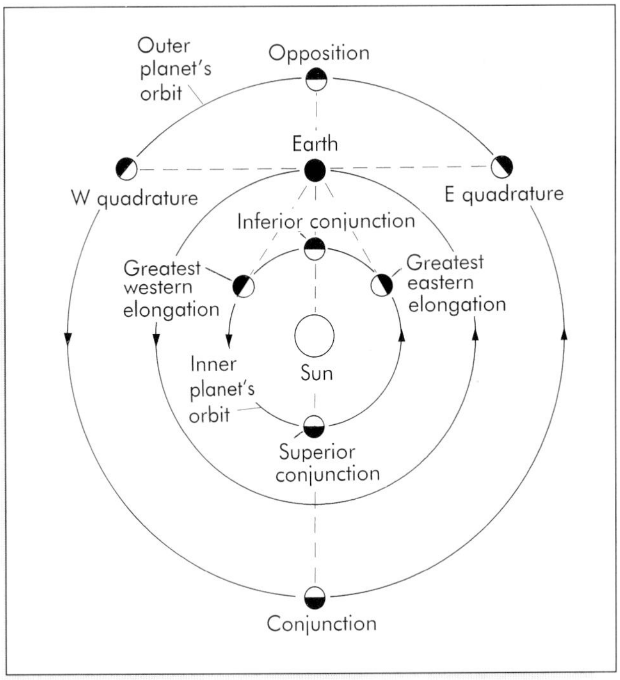

The aspect of planets – i.e. the relative geometrical alignments of the Earth, Sun and planets.

aspheric

A non-spherical surface for an optical component. The term encompasses the parabolic and hyperbolic shapes for the mirrors of *reflecting telescopes*, but is more commonly used for lens surfaces, since most lenses have spherical surfaces.

Association of Universities for Research in Astronomy

A consortium of 29 US universities and 6 international associates that provides major observing facilities including the *Space Telescope Science Institute*, telescopes at *Kitt Peak* and *Cerro Tololo*, the *National Solar Observatory*, and the *Gemini telescopes.* See http://www.aura-astronomy.org/h/hBot.html.

asterism

1 A recognisable collection of bright stars that is not a constellation, such as the seven stars that form the Plough (or Big Dipper) within the constellation of Ursa Major.
2 A star cluster with only a few member stars.

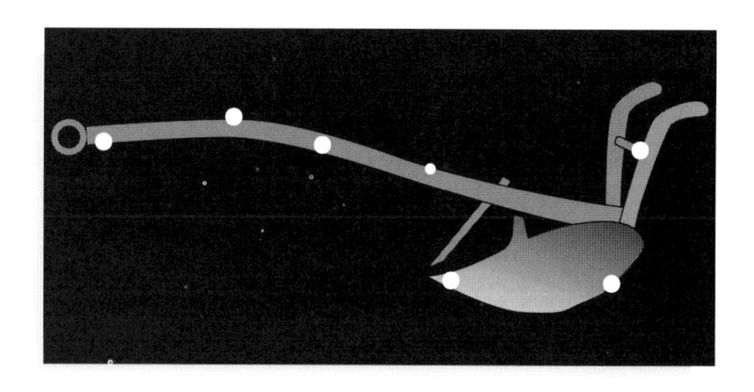

The asterism known as the Plough – the constellation of Ursa Major comprises these seven stars plus many more faint ones.

astigmatism

An *aberration* of images caused by a light beam in one plane coming to a different focus from a light beam in the orthogonal plane. The effect is most easily pictured for a cylindrical lens or mirror where the light in the plane of the axis of the lens or mirror is unaffected (i.e. its focus is at infinity), while that at right angles is brought to a nearby focus.

astrograph

An *optical telescope* optimized for determining the positions of objects in the sky. Astrographs are usually long focus *refractors* that can image a comparatively wide area of the sky and mostly date from nineteenth and early twentieth centuries. See also: *astrometry.*

astrolabe

A device for measuring the *altitude* of an object in the sky. Modern versions of the instrument, such as the *Danjon* or impersonal astrolabe, use a bath of mercury in order to give a precise horizon. From ancient times to at least the seventeenth century, however, an astrolabe comprised (in its most basic form) a circular plate, with a sighting bar (the alidade) pivoted at its center. The plate was suspended from a ring in order to give the vertical, and the alidade rotated to point at the object. The altitude of the object could then be read off the scale around the edge of the plate. The astrolabe was used to determine time at night, and as an aid to navigation. See also: *almucantar.*

astrometric eyepiece

An *eyepiece* with an illuminated internal scale (or *graticule*) that may be used to measure binary star separations, lunar crater diameters, etc. It may also incorporate an angular scale so that position angles may be found as well. See also: *bi-filar micrometer; cross wire eyepiece.*

astrometry

The science of the precise measurement of the positions of objects in the sky. Until recently, *astrographs* and *transit telescopes* were the main instruments used to determine the positions. Now, the *photographic zenith tube*, the *Danjon astrolabe*, and the *Hipparcos spacecraft* are the main source of precise positional measurements.

Astronomical Almanac

See *almanac.*

Astronomical Data Center

A data center for astronomical information maintained by NASA at the Goddard Space Flight Center. It may be contacted via http://adc.gsfc.nasa.gov/. See also: *Centre de Données astronomiques de Strasbourg; National Space Science Data Center.*

Astronomical Journal
A USA-based astronomy research journal published by the *American Astronomical Society*.

astronomical latitude
The direction between a plumb line and the equator. It equals the geocentric latitude, ϕ, at sea level, and then increases with height due to centrifugal 'force' arising from the Earth's rotation by an amount, $\Delta\phi \approx 0.00017h \sin 2\phi$, where h is the altitude above sea level in meters.

astronomical refractor
See *refracting telescope*.

astronomical societies
See list in the Appendix.

Astronomical Society of the Pacific
A society for amateur and professional astronomers based in western USA. See http://www.aspsky.org/.

astronomical telescope
Any telescope can be used for astronomical observation, but those producing *inverted images* are often labelled 'astronomical telescopes'. See also: *terrestrial telescope*.

astronomical twilight
See *twilight*.

astronomical unit
A non-SI measure of distance, that is convenient for use within the solar system, symbol, AU. It is defined as the length of the *semi-major axis* of the Earth's orbit and has a value of 1.49598×10^{11} m. See also: *light year; parsec*.

Astronomy
A popular USA-based astronomical magazine.

The popular astronomy magazine *Astronomy*.

Astronomy and Astrophysics
A Europe-based astronomy research journal.

Astronomy and Geophysics

A UK-based popular astronomy journal published by the *Royal Astronomical Society* for its members.

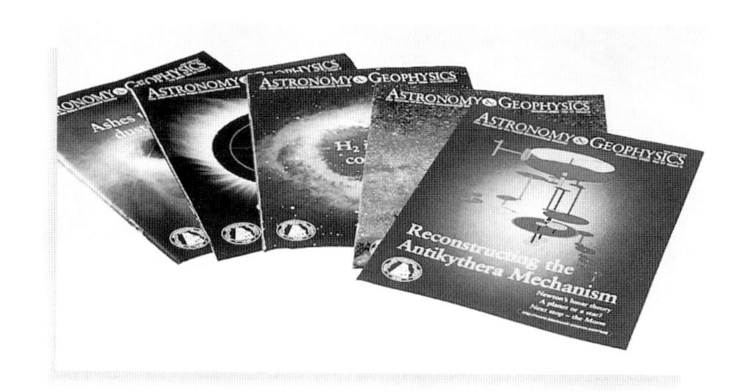

The popular astronomy magazine *Astronomy and Geophysics*.

Astronomy Now

A popular UK-based astronomical magazine. See http://www.demon.co.uk/astronow.

The popular astronomy magazine *Astronomy Now*.

Astrophysical Journal

A USA-based astronomy research journal published by the *American Astronomical Society*.

Astro-Physics

A commercial firm manufacturing telescopes and accessories.

Atacama large millimeter array

A planned *array* of sixty-four 12m dish type telescopes to observe at millimeter wavelengths, sited at Llano de Chajnantor, Chile, altitude 5,000 m. See homepage at http://www.alma.nrao.edu/info/index.html.

A concept image of the Atacama large millimeter array. (Image reproduced by courtesy of the National Radio Astronomy Observatory, Associated Universities Incorporated and the European Southern Observatory.)

atlas

See *star chart*.

atmospheric absorption

See *atmospheric extinction.*

atmospheric extinction

The reduction in the brightnesses of objects due to absorption in the Earth's atmosphere. The extinction varies with the observing conditions, *wavelength*, height of the observing site, and *zenith distance* of the object being observed. For a good site, the absorption at the zenith may vary from around 0.5^m (35%) in the violet to 0.05^m (5%) in the deep red. For accurate *photometric* work, the atmospheric extinction must be measured and the results corrected for its effects to give the magnitude of the object as if it were at the zenith. For zenith distances up to about 60°, this may be accomplished using Bouguer's law:

$$m_{\lambda,0} = m_{\lambda,z} - a_\lambda \sec z,$$

where $m_{\lambda,0}$ is the magnitude at wavelength, λ, and at the zenith, $m_{\lambda,z}$ is the magnitude at wavelength, λ, and at zenith distance, z, and a_λ is the extinction coefficient at wavelength λ. a_λ must be measured anew on every observing occasion by observing one or more *standard stars*. At larger zenith distances, the overall extinction increases from 0.3^m at z = 70° through 1^m at z = 80° to 4^m or more on the horizon. See also: *air path.*

atmospheric refraction

The change in the position of objects in the sky towards the zenith arising from the bending of light paths (refraction) within the Earth's atmosphere. The movement towards the zenith is given approximately by 58.2" tan z for *zenith distances* (z) up to about 60°. On the horizon, refraction moves objects upwards by about half a degree.

atmospheric window

A region of the *spectrum* over which *atmospheric extinction* is low, enabling radiation to penetrate to the surface of the Earth. The main windows are in the *optical region* (near *ultraviolet* to *infrared*) and the *radio region.*

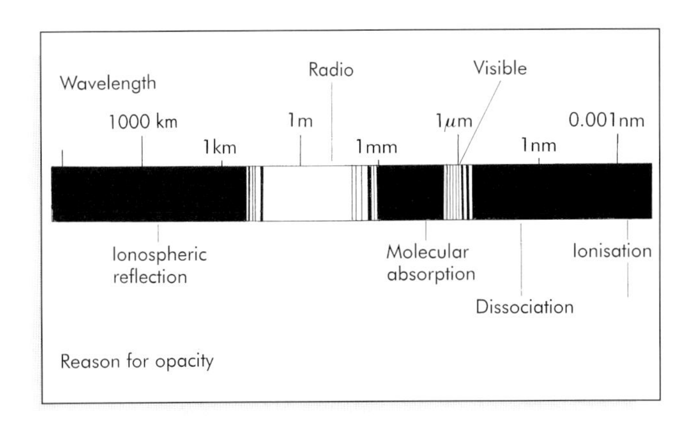

The wavelengths (windows) for which the Earth's atmosphere is transparent.

attenuation

See *atmospheric extinction* and *interstellar absorption.*

AURA

See *Association of Universities for Research in Astronomy.*

A

aurora

A phenomenon within the Earth's atmosphere, likely to be seen regularly by keen observers at high latitudes. It is a red or green glow that may appear as arcs, bands, rays, searchlights, curtains, etc. often with rapid movement and changes. The glow is due to emission from nitrogen and oxygen atoms that have been excited by high energy electrons from the Sun. The electrons arise from solar flares and coronal mass ejections, and so occur most frequently around the times of sunspot maximum. The auroras are concentrated in two annuli about 40° across and centered on the north and south magnetic poles (giving the northern lights or aurora borealis, and the southern lights or aurora australis), but during intense solar storms may extend to the southern UK and USA. A forecast of possible auroral activity may be obtained by registering by e-mail at aurorawatch-activity@samsun.york.ac.uk. See also: *airglow.*

The Aurora Borealis – 7 April 2000.

Australia telescope

A radio telescope with six 22m radio dishes in a 6km *array* at Culgoora in Australia. The array can be linked to dishes at Siding Springs and *Parkes* to give a maximum base-line of 300 km.

autocorrelation function

A measure of the degree of similarity between different sections of a signal, i.e. its degree of repetitiveness. The greater the repetitiveness, the higher the value of the autocorrelation function. It is given by:

$$\int_{-\infty}^{\infty} s(x)s(x-t)dx,$$

where $s(x)$ is the signal. It is a part of the data processing required to obtain the *spectrum* of a radio source.

autocorrelator

A device used as a part of the processing of a radio signal to obtain the *spectrum* of the object being observed. See also: *correlator.*

autoguider

A system for automatically keeping a telescope accurately aligned on the object being observed, used especially during imaging requiring long exposures. Many *CCDs* now incorporate an autoguiding system such as *TAA* or *STAR* that use the imaging signal itself to generate corrections to the drives. Alternatively, the autoguider may be a separate device, perhaps attached to a separate telescope, or sampling the beam from the main telescope, and using a detector such as a pin *photodiode* or a *photomultiplier* to detect when the image drifts. The system will then generate a correction signal for the telescope drives. See also: *guiding.*

automatic telescope

See *robotic telescope.*

autumnal equinox

The point in the sky where the *ecliptic* and the *celestial equator* cross and where the Sun's annual motion takes it from north to south of the equator. The Sun passes through it on or about 21 September every year. See also: *First Point of Aries.*

avalanche photo-diode

See *photo-diode.*

averted vision

The practice of looking slightly to the side of a faint object in order to see it more clearly. Averted vision gives a clearer view because the image then falls onto a part of the eye's retina that is rich in rod cells (which provide monochromatic vision). When one looks directly at an object the image falls onto the *fovea centralis*, a region of the retina which contains mostly cone cells (which produce color vision). Since rod cells are about 100 times more sensitive than cone cells, the faint object thus becomes much more easily seen when it is not looked at directly. See also: *dark adaptation; retina.*

Axiom Research

A commercial firm manufacturing *CCD* cameras for astronomical use.

azimuth

The angle of an object in the sky around from the north point of the horizon, symbol, A. It is measured from 0° to 180° East or West. It is related to the *RA* (α), *Dec* (δ), of the object, the observer's latitude (ϕ) and the local *sidereal time* (LST) by:

$$\sin A = \frac{-\cos\delta \sin(LST - \alpha)}{\sqrt{1 - (\sin\delta \sin\phi + \cos\delta \cos(LST - \alpha)\cos\phi)^2}}.$$

Note that LST and RA must be converted to degrees before inserting into this formula. See also: *altitude.*

B

BAA
See *British Astronomical Association.*

background noise
See *noise.*

background star
See *field star.*

background subtraction
The removal of background *noise* from a signal. See also: *dark exposure.*

back illuminated CCD
See *CCD.*

backlash
A delay occurring in the operation of a mechanical system when its action is reversed. Backlash arises from the 'play' that must be allowed between gears and other moving mechanical parts if excessive wear is to be avoided. It is often noticeable on a telescope's declination drives, where there will be a time lag before an adjustment in position occurs, after the telescope has been moving in the other direction.

baffle
A component within an optical system whose purpose is to shield the image from unwanted light. For example many *Schmidt-Cassegrain telescopes* have a tube projecting from the center of their *primary mirrors* that prevents light reaching the *eyepiece* directly from the sky around the sides of the *secondary mirror.* Many eyepiece designs incorporate *stops* that act in a similar manner.

Baily's beads
These are seen just before a solar eclipse becomes total. They are the last portions of the photosphere, glimpsed through valleys on the edge of the Moon. (CAUTION – always take appropriate precautions when observing the Sun or damage to the eye and/or telescope can result.) See also: *diamond ring.*

Baker–Schmidt camera
A variation on the *Schmidt camera* that has a flat focal surface through the use of a secondary mirror.

ball and claw mounting
A mounting for a small telescope that comprises a small sphere on a support and a clamp that can be loosened to allow the telescope to be moved to any part of the sky. This type of mounting is generally unsatisfactory and should be avoided. See also: *alt–az telescope mounting; equatorial telescope mounting.*

ball mounting

A mounting suitable for small reflecting telescopes, in which the bottom end of the telescope tube comprises part of a smooth sphere on a simple three-point support, and thus able to swivel in any direction.

The ball telescope mounting.

B

Balmer discontinuity

A sudden drop in the intensity of the emission from stars that starts at 364.6 nm and continues to shorter wavelengths, due to *ionization* of hydrogen atoms that are already *excited* to their first energy level. See also: *Balmer lines.*

Balmer jump

A synonym for *Balmer discontinuity.*

Balmer lines

The *spectrum lines* of the hydrogen atom in the *visible* part of the spectrum. They arise from *transitions* from the first excited level to higher levels, or vice versa. The lines are labelled with Greek letters starting with H-α at 656.3 nm and continuing with H-β (486.1 nm), H-γ (434.0 nm), H-δ (410.1 nm), etc. The lines continue getting closer in wavelength terms and generally become weaker until the series limit is reached at 364.6 nm. Similar series of lines arise in the *ultraviolet* due to transitions to and from the ground state (Lyman series), and in the *infrared* (Paschen series, transitions to and from the second excited level, Brackett series, transitions to and from the third excited level, etc.). See also: *Fraunhofer lines.*

balun

A transformer used to connect a balanced *feed* for a *radio telescope* (such as a *half-wave dipole*) to an unbalanced transmission line (such as coaxial cable).

band pass filter

A *filter* that transmits a band of *wavelengths*, and absorbs or rejects wavelengths outside that range. Also an electronic device that only allows through signals over a restricted frequency range. See also: *filter.*

bandwidth

The range of *wavelengths* or *frequencies* occupied by a signal, transmitted by a *filter* or over which an item of electronic equipment responds.

B

Barlow lens

A negative or *diverging lens* placed in front of an *eyepiece* in order to increase the *effective focal length* of the telescope, and so to increase the magnification of the eyepiece. Some manufacturers incorporate Barlow lenses into their eyepieces, so that they have the same *eye relief* irrespective of their focal length.

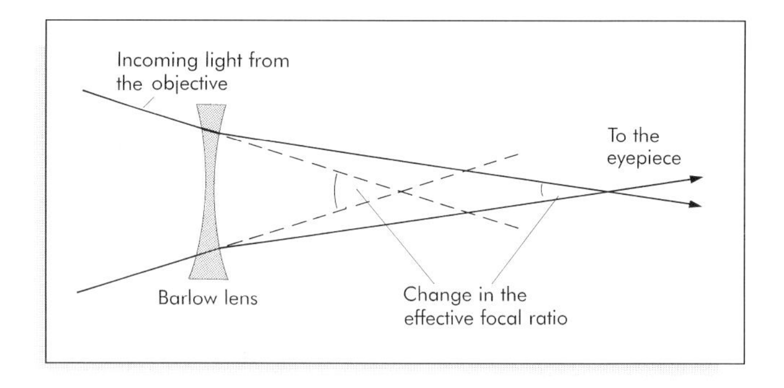

The optical principles of the Barlow lens.

barn door mounting

A mounting for a camera that enables it to track objects in the sky. It may simply and cheaply be made by a DIY enthusiast. It comprises two flat boards that are joined by a hinge along one of their sides. The hinge axis is aligned parallel to the Earth's rotation axis (i.e. on the north or south point in the sky). The camera is mounted on the upper board, and the tracking movement produced by a bolt through the lower board that bears on the upper board. The bolt is turned by hand at intervals of a few seconds. The amount and rate of motion required for the bolt will depend upon its thread angle and the distance from the hinge, and can be calculated or found by trial and error. The device is best suited to cameras with short focal lengths. Synonyms are the Haig mount and the scotch mount.

barrel distortion

See *distortion.*

base line

The separation of two components of an *interferometer.*

Bayer name

The system of labeling stars by a Greek letter in the order of their brightnesses within the *constellation* followed by the constellation name. For example α Ori, γ Cyg, or ε Eri. The system was introduced by Johan Bayer (1572–1625) in the Uranometria star catalog of 1603. There are often anomalies to the system arising from changes to the constellations or mistakes, with some constellations missing some letters, or brighter stars having later letters than fainter ones, etc. After the Greek letters have been used, then the lower case letters, a, b, c, . . ., may be used in a similar way, and after the lower case letters, upper case letters from A to P. The letters from R to Z are reserved for variable stars (see *variable star nomenclature*). See also: *Flamsteed number; stellar nomenclature.*

BD catalog

See *Bonner Durchmusterung catalog.*

beam efficiency

The ratio of the power in the *main lobe* of a radio transmitter to the total power being emitted. See also: *aperture efficiency.*

beam splitter

An optical component that splits an incoming beam of radiation into two separate beams. It is usually a partially reflecting plane mirror. See also: *dichroic mirror; Nicol prism; Wollaston prism.*

beam width

A measure of the angular *resolution* of a *radio telescope* or *interferometer*. It is normally taken as the angle between the *half-power points* of the *main lobe* of the instrument. It is a smaller measure of resolution than would be given by the *Rayleigh criterion*, which would be the angle between the first nulls on either side of the main lobe.

Beijing observatory

An optical observatory, based in Beijing, China, and with several observing stations. Its largest instrument is a 2.2m reflector.

Besselian day numbers

A set of five numbers that enable the position of an object in the sky to be corrected for the effects of *precession*, *nutation* and *aberration*. They are tabulated for each day in the Astronomical *Almanac*. See also: *independent day numbers*.

best fit

See *least squares fit*.

Beral

A propriety metallic reflective coating for mirrors. It is hard enough to withstand cleaning and has around 90% reflectivity from about 250 nm to well into the infrared.

bi-concave lens

A lens with both surfaces concave in shape. See also: *meniscus lens*.

bi-convex lens

A lens with both surfaces convex in shape. See also: *meniscus lens*.

bi-filar micrometer

An eyepiece that has three fine threads mounted at its internal focal plane. Two of the threads are orthogonal and fixed, as in the *cross wire eyepiece*. The third is parallel to one of the fixed threads and may be moved by means of a calibrated screw along the direction of the other thread. The whole eyepiece may be rotated and its orientation read off from a scale. The threads can be focused separately from focusing on the image, so that both can be seen sharply. The threads can be illuminated so that they may be seen against a black background. Spider webs were commonly used for the threads at one time, but that has now largely been replaced by thread obtained by unravelling dental floss. The bi-filar micrometer is used to measure small angles, such as the diameters of planets or the separation of double stars, and *position angles*. However, it has now mostly been superseded by direct measurement of these quantities on *CCD* images. See also: *astrometric eyepiece*.

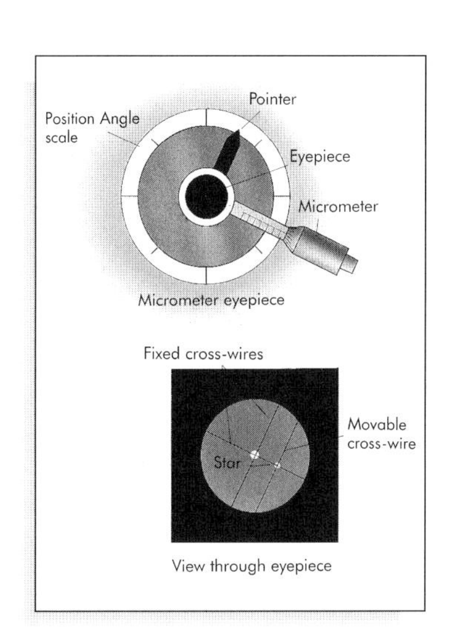

A bifilar micrometer eyepiece.

Big Bear solar observatory

An observatory in the San Bernardino mountains California, altitude 2100 m, that houses a *tower solar telescope* surrounded by a lake in order to obtain stable atmospheric conditions.

binary star

A *double star* in which the two components are relatively close to each other in space and are linked gravitationally.

B

binning
Combining the outputs from several *CCD pixels* to give an image with lower noise but also lower resolution.

binocular eyepiece
A device that enables both eyes to be used to view an image through a telescope. It replaces the conventional eyepiece and contains a beam splitter and mirrors to direct two beams of light into separate matched eyepieces. One eyepiece will usually have a separate means of focusing and the distance between the eyepieces may be adjustable. The device does not provide stereoscopic vision (except as an optical illusion) since both eyepieces receive light from the single light path entering the telescope. However, many people find that observing with both eyes is easier and less tiring than using one eye.

binoculars
A pair of small telescopes, usually providing an erect image, joined together so that both eyes may be used to view an object simultaneously. The physical length of the binocular is often reduced by internal prisms, which reflect the light paths back and forth. Binoculars are specified by the *magnification* and *aperture*, for example, 7 × 50 binoculars have a magnification of ×7, and *objectives* with diameters of 50 mm. Binoculars are usually used for viewing terrestrial objects. They are often recommended as an introductory-level instrument for budding astronomers, but are likely to be disappointing since their magnifications will usually be too low to show much. Giant binoculars (e.g. 20 × 100, etc.), however, are available which do give magnificent views of the heavens, though they are more expensive than an equivalent telescope.

Binoculars.

Binoculars in which the image is stabilized (against vibrations and wobbles from being hand-held) have recently come onto the market. They use sensors for small movements of the binoculars and a computer-controled compensating optical system to keep the image steady. With these, hand-held binoculars with magnifications up to about ×25 are practical.

birefringence
The property of some crystals and other materials whereby beams of linearly polarized light behave differently depending upon their orientation to the crystal structure. It is usually possible to find one orientation of the plane of polarization for which the material behaves in a 'normal' optical fashion. This is called the ordinary ray. Light polarized at right angles to the ordinary ray will have a *refractive index* that varies depending upon the direction of passage of the ray through the crystal. This is called the extraordinary ray. The optical axis of the crystal is the direction where the refractive indices of the two rays are the same. Birefringent materials are used in the *Lyot filter*, *polarizers* and *wave plates*.

birefringent filter
See *Lyot filter*.

BIS
See *British Interplanetary Society*.

black and white film

Photographic film with a single type of *photographic emulsion* that produces images in various shades of grey ranging from black to white (sometimes called a monochromatic film).

B

black body

See *black body temperature.*

black body temperature

A black body is an object that absorbs perfectly at all wavelengths. A good practical realization of it is as a small hole in a box. When a black body is heated it radiates according to Planck's equation (see *thermal radiation*), with the black body temperature, T, and the wavelength at which the emission is a maximum, λ_{max}, given by Wien's law: $\lambda_{max}T = 0.0028979\ mK$. The total energy emitted per unit area by a black body, known as the flux, F, is related to the temperature by the Stefan Boltzmann law: $F = 5.6696 \times 10^{-8}T^4\ Wm^{-2}$.

black drop

See *tear drop effect.*

blank

See *mirror blank.*

blazed grating

A *diffraction grating* with the individual elements ruled at an angle in order to reflect most of the energy away from the zero order and into a higher order. By blazing a grating up to 90% of the energy can be channelled into the desired *spectral order.*

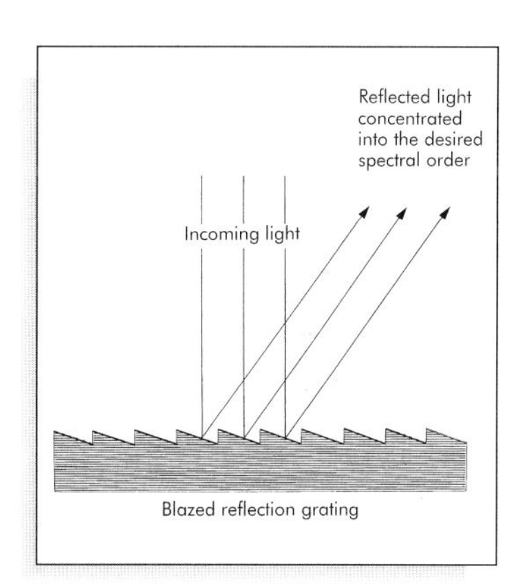

A highly magnified cross-section through a blazed diffraction grating.

blink comparator

A device that alternately displays two images of the same area of sky obtained at different times at intervals of a second or so. When the images are correctly aligned, stationary and constant stars will be flicker-free, but a star that has changed in brightness will blink, so calling attention to itself and greatly easing the discovery of variable stars. Likewise an object that has moved will jump back and forth as the images are alternated. Early comparators physically alternated the view of photographic images, but nowadays it is usual for the images to be displayed on a computer monitor, and software used to alternate the views.

blocking filter

A synonym for *cut-off filter.*

blooming

1 A coating, for example of magnesium fluoride, on the surfaces of a lens which cuts the amount of light reflected from the surfaces by about a factor of three when compared with an uncoated lens. See also: *multi-coated optics.*

2 An overflow of electrons from a saturated pixel within a CCD detector into those surrounding it. On the image the effect is for very bright objects to have bright lines extending away from them. See also: *anti-blooming*.

blue moon

Now commonly taken to be the second full moon occurring within a single calendar month. Such an event happens about once every three years. The next few blue moons will be in October or November 2001 (depending on geographical location), July 2004, May 2007, December 2009, August 2012, and July 2015.

The traditional definition is that it is the third full moon in a season that has four full moons within it. On that basis the next few blue moons occur in November 2002, August 2005, May 2008, November 2010, August 2013 and May 2016.

Very occasionally, the Moon may actually appear blue in color due to light scattered by dust within the Earth's atmosphere.

blue shift

See *Doppler shift*.

magnitude

See *UBV photometric system*.

bolometer

A detector that changes its electrical resistance in response to heating by the illuminating radiation. Liquid helium cooled semi-conductor bolometers are used as detectors in the infrared, and are made from beryllium or gallium-doped germanium. See also: *photoconductive cell*.

bolometric correction

The difference between the *bolometric magnitude* and the *V magnitude* (and sometimes vice versa). Its value is zero for stars of spectral type F5, –4 for B0 stars, and –3 for M5 stars. The bolometric correction is obtained from a combination of observation and theory. It is used to obtain the bolometric magnitude of a star from its V magnitude.

bolometric magnitude

The *magnitude* of a star based upon its total energy emission at all wavelengths. It is obtained from the *V magnitude* using the *bolometric correction*.

Bolshoi telescop azimutalnyi

A 6m *Ritchey–Chrétien* optical telescope using a monolithic mirror on Mount Pastukhov in the Caucasus. See also: *Zelenchukskaya observatory*.

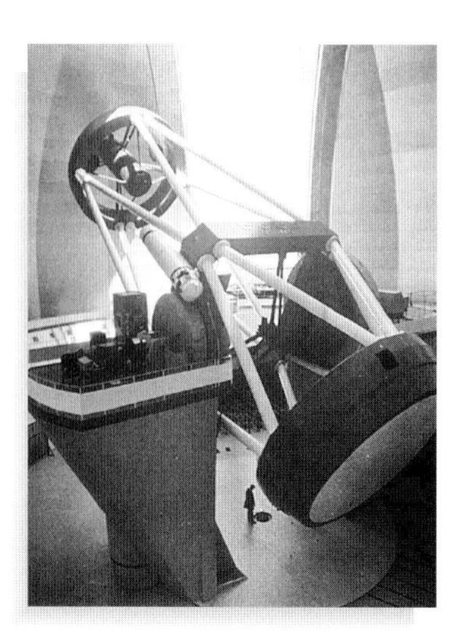

The Russian 6m (236-inch) telescope named the Bolshoi telescop azimutalnyi.

Boltzmann's equation
See *excitation temperature.*

B

Bonner Durchmusterung catalog
A star catalog containing some 450,000 star positions with a limiting magnitude of $+9.5^m$, and published between 1859 and 1887. It only covers the sky as far south as *declination* −23°. Nonetheless many stars are still known by their BD numbers, which take the form of the star's declination, and a running number, e.g. BD +12° 1234. The Córdoba Durchmusterung (CD) catalog published in 1930, completes the southern coverage and contains over 600,000 stars to a limiting magnitude of $+10^m$.

Bouguer's law
See *atmospheric extinction.*

bound-free transition
See *ionization.*

Bouwers telescope
See *Maksutov telescope.*

Brackett series
See *Balmer series.*

Bragg spectrometer
A *spectrometer* operating in the *X-ray region* that uses a crystal of lithium fluoride, graphite, etc. to separate the X-rays according to their *wavelengths.*

bremsstrahlung radiation
See *free-free radiation.*

Bresser
A commercial firm manufacturing telescopes and accessories.

bright star catalog
A catalog of the properties of stars brighter than 6.5^m. It contains over 9,000 entries and was first published in 1930 by the Yale university observatory. See also: *Harvard revised catalog.*

brightness temperature
See *effective temperature.*

British Astronomical Association
A UK-based society for amateur and professional astronomers. See http://www.ast.cam.ac.uk/~baa/.

The British Astronomical Association's Handbook.

B

British Interplanetary Society
A London-based society for supporting and promoting interest in space and astronautics. See
http://.bis-spaceflight.com/homepage.htm.

broad band photometry
Photometry using *filters* with *bandwidths* of 100 nm or more. The *UBV system* is the best-known example.

broadside array
See *array*.

BTA
See *Bolshoi telescop azimutalnyi*.

Bushnell
A commercial firm manufacturing telescopes and accessories.

Butler matrix
A system of connecting together the elements of an *array* or *interferometer* so that the transmission line lengths are the same for all elements. Pairs of elements are first connected together, then pairs of pairs, then pairs of pairs of pairs, and so on.

B–V color index
See *color index*.

Byurakan astrophysical observatory
An optical observatory on Mount Aragatz, Armenia, altitude 1400 m. It houses a 2.6m telescope.

C

C band
An X-ray band covering the wavelengths 5 nm to 10 nm.

caesium iodide detector
See *scintillation detector*.

Calar Alto observatory
An observatory housing 3.5m, 2.2m and other large optical telescopes located on Calar Alto, Andalusia, altitude 2200 m.

Caldwell catalog
A catalog of 109 interesting objects for observing with small telescopes, developed in 1995 by Sir Patrick Caldwell Moore (1923–). The catalog mirrors the *Messier catalog* that has a similar number of objects within it. The Caldwell catalog, however, covers the whole sky and the running numbers increase consistently from north to south. Objects are designated by C plus the running number, for example the star cluster h and χ Per is C14. See Appendix for full list.

calendar
See *Gregorian calendar* and *Julian calendar*.

calibration
A process of setting an instrument to give accurate results by measuring an object with known properties. The instrument is either adjusted to give the correct reading, or a table drawn up to correct the *raw data*.

calibration exposure
A photograph of a number of sources of differing but known relative intensities, used to produce the *characteristic curve* of the *photographic emulsion*.

Caltech sub-millimeter observatory
A sub-millimeter wave observatory, with a 10.4m dish, on *Mauna Kea*, Hawaii.

Cambridge optical aperture synthesis telescope
An optical *aperture synthesis* system using four 0.4m mirrors being developed at Cambridge (UK).

camcorder
See *video camera*.

camera
See panel (pages 32–33).

Canada balsam
A sticky transparent liquid obtained by heating the sap of Canada balsam trees in order to evaporate some of the natural turpentine. It is used to cement lenses together.

camera

Any self-contained device for producing an image. The term is usually applied to devices used in *photography*, with *CCD detectors* or for TV or *video*. The basic components of a camera are a lens whose position may be changed to permit focusing, a variable speed shutter to provide a range of exposures, and the detector. Almost all cameras, however, will also have a variable diaphragm to change the *focal ratio*, an exposure meter, and a viewfinder. Often the lens may be exchanged, or a zoom lens used, so that the *image scale* may be varied. Automatic exposure and focusing is common on modern cameras, but these devices may not work if the camera is used on a telescope. For astronomical purposes therefore a simple camera may be better. See also: *single lens reflex camera; digital camera; video camera.*

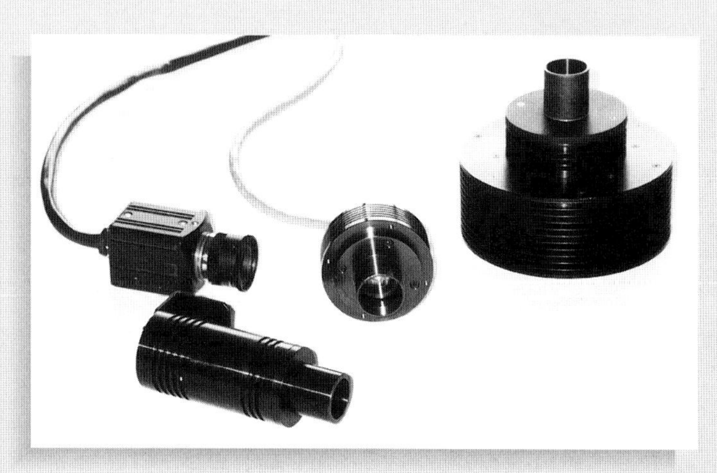

CCD cameras for astronomical imaging.

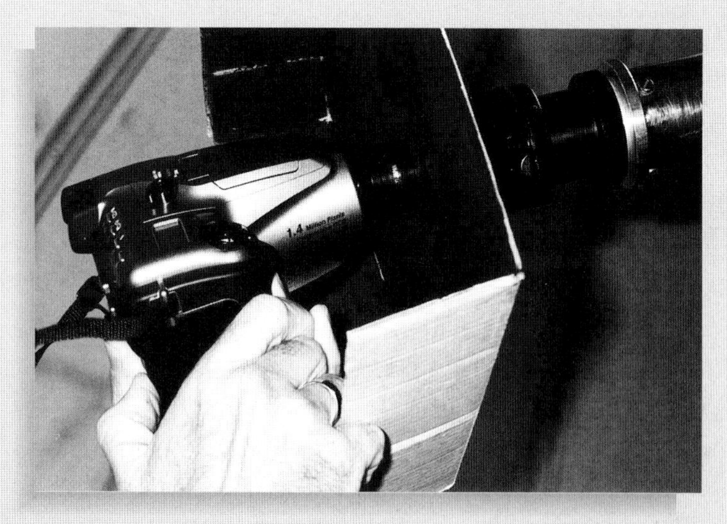

A digital still camera being used for afocal imaging.

A photographic SLR camera being used with an off axis guider for imaging at the Cassegrain focus of the telescope. (Image reproduced by courtesy of Meade Instruments Corp.)

A video camera being used for afocal imaging on a Newtonian telescope.

A battery of fixed photographic cameras for imaging meteors.

Canada–France–Hawaii telescope

A 3.6m optical and infrared telescope located on *Mauna Kea*, Hawaii that has pioneered many of the methods used to improve image quality.

C

candela

The SI unit of luminous intensity.

cardinal points

See *compass points.*

Carrington rotation

A system for identifying rotations of the Sun developed by Richard Carrington (1826–75). The Carrington rotations start at 12:00 UT on the 1 January 1854. The solar rotation period increases with solar latitude and so the rotations are based on the solar *sidereal rotation period* at the solar latitudes ±20°, which is 25.38 days. The Carrington number increases by one every *synodic rotation* period of the Sun, which averages 27.2753 days. The start of a Carrington rotation corresponds to 0° of solar longitude at ±20° of solar latitude crossing the center line of the Sun's disk as seen from the Earth. Carrington rotation number 1958 started at 14h 40min UT on 1 January 2000.

Carte du Ciel

A photographic atlas of the sky with a *limiting magnitude* of 14^m, conceived at the end of the nineteenth century, but never fully completed.

Cassegrain telescope

See panel (page 35).

Cassegrain–Maksutov telescope

See *Maksutov telescope.*

catadioptric telescope

A telescope that uses both a lens and a mirror to gather the light. The combination can result in a very compact design with excellent images over a comparatively wide *field of view*. Many small instruments are produced to designs of this type for the amateur market. See also: *Schmidt–Cassegrain telescope; Maksutov telescope; ETX telescope.*

catalog

See *star catalog.*

catalog equinox

The zero point for *right ascension* of a star catalog. It will be valid for the *epoch* of the catalog, but will differ from the actual zero point, or *dynamical equinox*, at any other instant due to *precession.*

Catalog of Galaxies and Clusters of Galaxies

See *Zwicky catalog.*

catoptric system

An optical system based only upon mirrors.

Cassegrain telescope

A design of *reflecting telescope* invented by Guillaume Cassegrain in 1672. It uses a parabolic *primary mirror* and a hyperbolic *secondary mirror*. The focus is at the bottom of the telescope tube with the light passing through a hole in the primary mirror. Most modern large telescopes are of the Cassegrain design, or its variant the *Ritchey–Chrétien* design. See also: *Newtonian telescope; Gregorian telescope.*

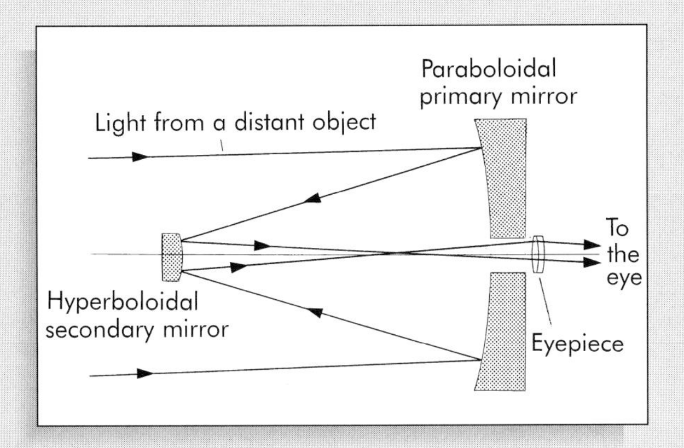

The optical principles of a Cassegrain telescope.

A 0.4 m (16-inch) Cassegrain telescope with a small spectroscope at the Cassegrain focus (Image reproduced by courtesy of Bob Forrest, University of Hertfordshire Observatory).

The Gemini North telescope – the mirror cell for the Cassegrain secondary mirror can be seen at the top of the telescope. (Image copyright 1999 Neelon Crawford – Polar Fine Arts, courtesy of Gemini Observatory & National Science Foundation).

CAT telescope

A synonym for a *catadioptric telescope* (the name comes from an abbreviation of "catadioptric").

C

Cauchy formula

An approximate formula for the *refractive index* of an optical material at wavelength, λ, of the form:

$$\mu_\lambda \approx A + \frac{B}{\lambda^2} + \frac{C}{\lambda^4} + \ldots,$$

where A, B, C, ... are constants for the material. See also: *Hartmann formula; constringence.*

CBAT

See *Central Bureau for Astronomical Telegrams.*

CCD detector

See panel (pages 37–38).

celestial equator

A great circle on the *celestial sphere* dividing it into northern and southern hemispheres. The celestial equator is the extension into space of the plane of the Earth's equator.

celestial latitude

A coordinate system, along with *celestial longitude*, for giving the positions of objects in the sky. The celestial latitude (symbol, β) is the angle, in degrees, up or down from the *ecliptic*, going from 0° to ±90°. It is related to the *RA* (α), *Dec* (δ), of the object and the *obliquity of the ecliptic* (ε) by, $\sin \beta = \sin \delta \cos \varepsilon - \cos \delta \sin \alpha \sin \varepsilon$. Note that the RA must be converted to degrees before inserting into this formula.

The co-latitude is the angle from the pole of the ecliptic, i.e. $90° - \beta$. See also: *declination; right ascension; celestial sphere.*

celestial longitude

A coordinate system, along with *celestial latitude*, for giving the positions of objects in the sky. The celestial longitude (symbol, λ) is the angle, in degrees, around the *ecliptic* from the *First Point of Aries*, going from 0° to 360°. It is related to the *RA* (α), *Dec* (δ), of the object and the *obliquity of the ecliptic* (ε) by:

$$\cos \lambda = \frac{\cos \delta \cos \alpha}{\sqrt{1 - (\sin \delta \cos \varepsilon - \cos \delta \sin \alpha \sin \varepsilon)^2}}.$$

Note that the RA must be converted to degrees before inserting into this formula. See also: *declination; right ascension; celestial sphere.*

celestial poles

The positions on the *celestial sphere* about which the latter appears to rotate. They are where the line of the Earth's rotational axis meets the celestial sphere. The north celestial pole is currently marked by Polaris, the pole star, which is within a fraction of degree of the true pole, but there is no equivalent star to mark the south celestial pole.

CCD detector

The principal optical imaging detector used by all professional and many amateur astronomers. The initials stand for Charge Coupled Device. The advantages of the CCD detector are sensitivity (20 to 50 times more sensitive than *photographic emulsion*), low background *noise*, *linear response*, high *dynamic range* and the production of images in electronic form ready for computer processing and analysis. The disadvantages of CCDs are small size, the need to be cooled and high cost.

The detection mechanism in CCDs is electron-hole *pair production* in p-doped silicon. A positive electrode on top of, but insulated from, the silicon attracts and stores the electrons, and repels the holes. Many such units (or *pixels*) are stacked in linear or two-dimensional arrays to produce the imaging detector. At the end of the exposure the charge packets stored within each pixel are physically moved through the silicon to a set of read-out electrodes by changing voltages. The charge packets are then moved to the output of the device by cycling the voltages on the read-out electrodes (this is the 'charge coupling' stage). CCDs may be three, two or virtual phase, with three, two or one read-out electrodes per pixel respectively.

The processing of a CCD image normally requires the subtraction of a blank image of the same exposure, known as the dark exposure, to reduce the background *noise*, *flat fielding*, and correction of *cosmic ray* strikes on the detector.

The optical sensitivity of a CCD ranges from around 400 nm to 1100 nm, with a peak around 750 nm where the *quantum efficiency* may approach 90%. Sensitivity to short wavelengths may be improved and extended by applying a surface coating of a phosphorescent material. The CCD is also naturally sensitive to X-rays, and can act as a direct spectrometer, since its response is proportional to the energy of the X-ray.

Small ($\sim 500 \times 500$ pixel) CCDs are currently produced for about the same cost as a 125 mm *Schmidt–Cassegrain telescope* and are now widely used by amateur astronomers. For professional use, devices up to about $10{,}000 \times 10{,}000$ pixels are available. These latter are very expensive, and since they are usually thinned to about 10 μm, so that they can be illuminated from the back in order to improve their sensitivity, they are also fragile. In addition, processing such large images requires substantial computing power and specialist software, well beyond the capacity of currently available PCs. See also: *SBIG; Starlight Xpress.*

CCD cameras for astronomical use.

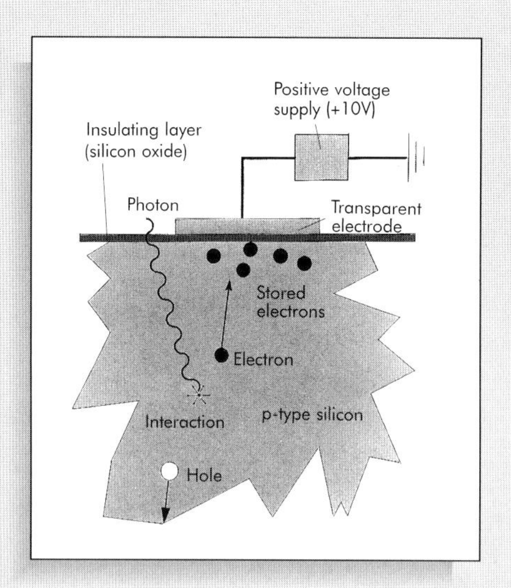

The structure of the basic unit of a CCD detector.

A CCD camera with a home-made filter wheel on a small telescope.

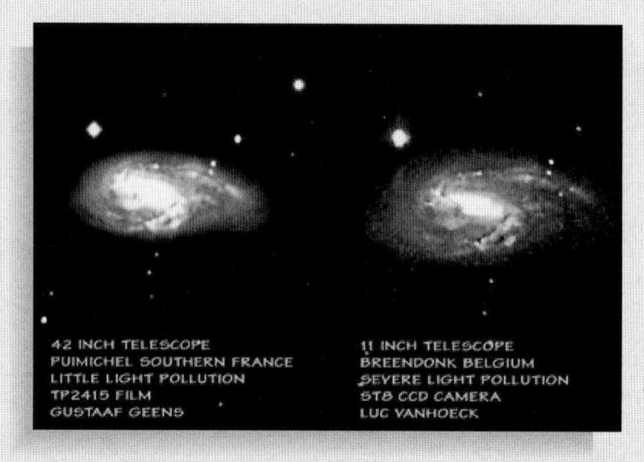

The power of CCD cameras – two 15 minute exposures of the galaxy M66 – a photographic image from a 1.07 m telescope on the left and a CCD image from a 0.28 m telescope on the right.

celestial sphere

An imaginary sphere, centered on the Earth, and large enough to contain all the objects in the universe. The positions of objects in the sky (e.g. *right ascension* and *declination*) are defined by taking a line from the center of the Earth, through the object, and then extending it until it meets the celestial sphere, to give a corresponding position on the sphere.

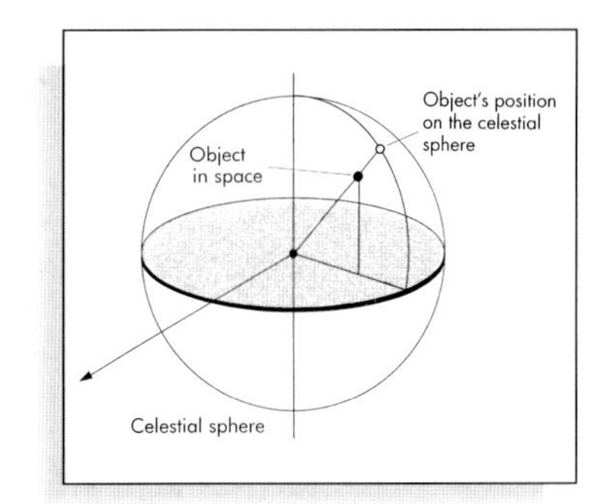

The celestial sphere.

C

Celestron

A commercial firm manufacturing telescopes and accessories.

cell

See *mirror cell.*

Central Bureau for Astronomical Telegrams

An organization set up by the *International Astronomical Union* and based at the Smithsonian Astrophysical Observatory to which discoveries of objects such as comets, novae, super novae, asteroids, etc. must be reported in order for you to be accepted by the astronomical community as the discoverer. It may be contacted if you think you have discovered something (but check very carefully first) at http://cfa-www.harvard.edu/cfa/ps/cbat.html or by searching the web for "CBAT".

central meridian

The imaginary line on the disk of a planet, satellite or the Sun joining the north and south poles, and dividing the object into its eastern and western hemispheres.

Centre de Données astronomiques de Strasbourg

The major center for collecting astronomical data, based at the Observatoire de Strasbourg, France. It may be contacted via; http://cdsweb.u-strasbg.fr/. The *SIMBAD* (Set of Identifications, Measurements and Bibliography for Astronomical Data) database, containing information on 2.8 million objects, may be accessed via the web on; http://cdsweb.u-strasbg.fr/Simbad.html. See also: *Astronomical Data Center; National Space Science Data Center.*

Center for High Angular Resolution Astronomy

An optical and infrared observatory on Mount Wilson, California. It operates a six-element array using 1 m telescopes that can reach resolutions of 0.0002".

Centre National d'Etudes Spatiales

A French organization overseeing space research and policy, including the French contribution to the *European Space Agency.*

Čerenkov detector

A device for detecting high-energy charged particles through their Čerenkov radiation. The latter occurs as a charged particle passes through an optical material at a speed greater than the local velocity of light (remember that *refractive index* is the ratio of the speed of light in a vacuum to that in the material, so that in glass, for example, light travels at about 200,000 km/s). Čerenkov radiation is analogous in terms of *electro-magnetic radiation* to the sonic boom produced by the breaking of the sound barrier. Čerenkov detectors are used to detect primary *cosmic ray* particles, through their emissions high in the Earth's atmosphere, and high-energy electrons produced by *neutrinos* in water-tank detectors such as *Kamiokande* and *IMB.*

C

Cerro Pachon observatory
An observatory housing the *Gemini South telescope*, sited on Cerro Pachon, Chile, altitude 2700 m.

Cerro Paranal observatory
An observatory housing the *very large telescope*, sited on Cerro Paranal, Chile, altitude 2600 m. See also: *European southern observatory.*

Cerro Tololo Inter-American observatory
An observatory housing several large optical telescopes including the *Victor Blanco telescope*, sited on Cerro Tololo, Chile, altitude 2200 m.

Cer-Vit
A low thermal expansion glass-ceramic material, widely used to form mirrors for large telescopes.

CFHT
See *Canada–France–Hawaii telescope.*

CHARA
See *Center for High Angular Resolution Astronomy.*

characteristic curve
A plot of the *optical density* of the image on a *photographic emulsion* against the logarithm of the exposure (i.e. intensity times exposure length). It is used to enable measurements of the relative brightnesses of objects to be made from photographic images.

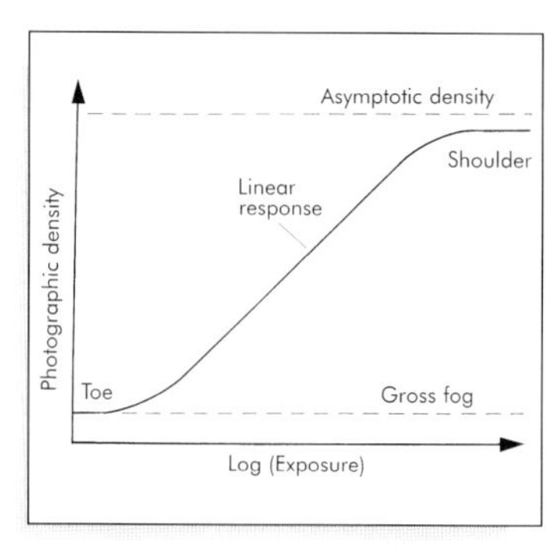

The characteristic curve of a photographic emulsion.

charge coupled device
See *CCD.*

chart
See *star chart.*

chi-squared test (χ^2 test)
A statistical test for the similarity of two distributions of data. It may be used to test hypotheses by using predictions as one set of data and observations as the other. See also: *significance level.*

chopping
A process used to reduce background *noise*, especially for infrared detectors. The detector is switched between the source and the background sky, a nearby standard source or an artificial standard source at frequencies between 1 kHz and 0.1 Hz. The chopping mechanism may take many forms such as an oscillating secondary mirror for the telescope or a rotating disk with clear and reflective segments that alternately permits through the light from the source or reflects the radiation from an artificial standard. Subtraction of the source and comparison signals together with other signal processing, such as integration, can allow the detection of sources a million times fainter than the background noise.

christmas tree

See *Butler matrix.*

chromatic aberration

A fault in images produced by optical systems that use lenses. It arises from the different values of the *refractive index* of the glass at different *wavelengths*. Light of different wavelengths is thus brought to different focal points resulting in colored fringes around the image of an object. It effects may be reduced by combining two or more lenses formed from different types of glass, and resulting in *achromatic* and *apochromatic lenses*. See also: *aberration (optical); chromatic difference of magnification.*

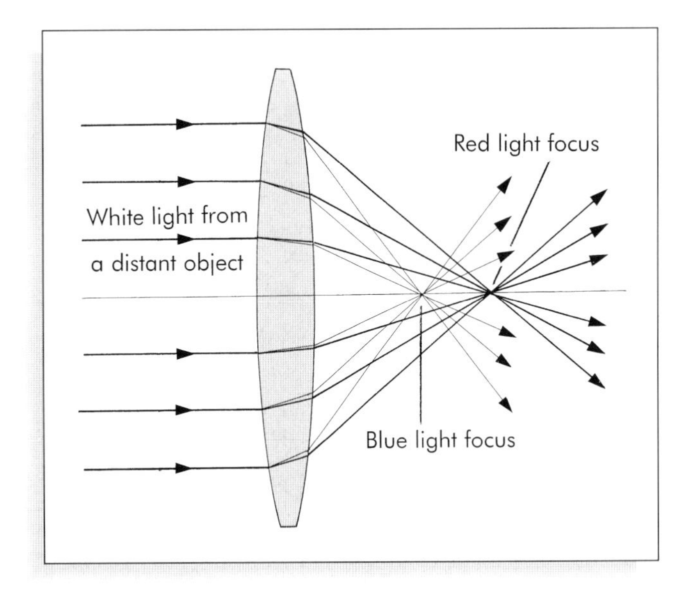

a Chromatic aberration for a simple lens.

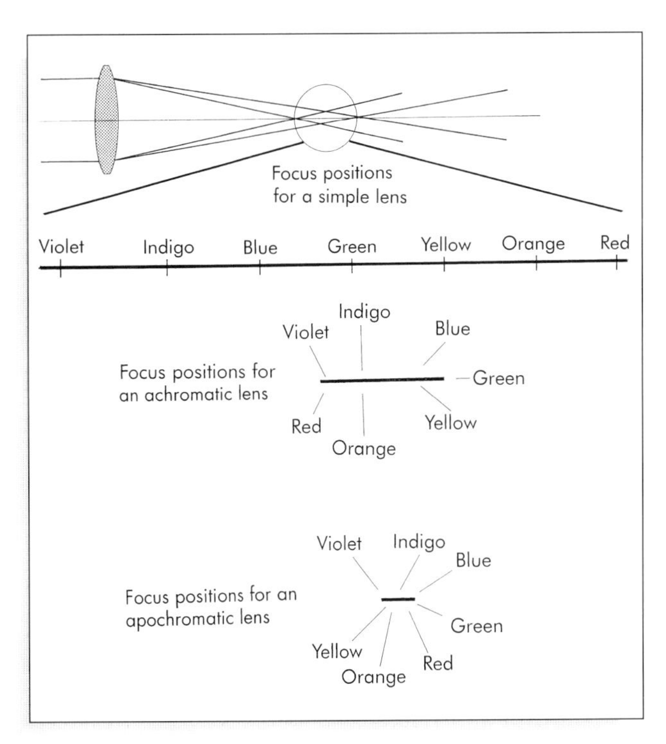

b A comparison of the chromatic aberration for a simple lens with the residual effects for an achromatic doublet and an apochromat.

chromatic difference of magnification

The difference in the *magnifications* of images at different wavelengths arising from the different *focal lengths* of the *lens* due to *chromatic aberration*, or to partially corrected chromatic aberration.

Ciel et Espace

A popular French-based astronomical magazine.

circle of least confusion

In an optical system that is affected by *aberrations*, this is the point at which the image appears to be in best focus.

circularly polarized radiation

Electromagnetic radiation in which the direction of the electric vector rotates with time at the frequency of the radiation. See also: *elliptically polarized radiation; linearly polarized radiation.*

C

circumpolar

An object that is either always above, or always below, the horizon. The term is mostly used for the former class of objects, which are thus visible at all times of the night and throughout the year. To be circumpolar, the angular distance of the object from the pole must be less than the observer's latitude.

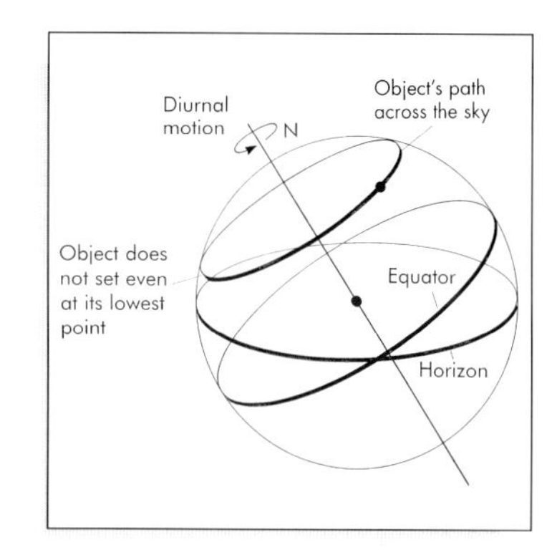

The path across the celestial sphere of a circumpolar object.

civil time

The time commonly used for every day life. It is based upon *universal time*, but changes according to the time zone, and often is advanced by one or two hours in the summer to give summer or daylight-saving time. See also: *coordinated universal time.*

civil twilight

See *twilight.*

Clairaut condition

A design for an *achromatic lens* in which the second and third lens surface match so that the two components may be cemented together to give a very rugged construction. See also: *Fraunhofer lens; Littrow condition.*

clam shell dome

A hemispherical *dome* for small telescopes that opens by means of two segments that pivot horizontally.

CLEAN

An *image processing* technique, mostly used on radio images, especially those from *aperture synthesis* systems. It is an iterative process. The most intense part of the *raw image* is found. A new image is produce by taking the *point spread function* for the instrument, centering it on that peak, and subtracting it from the image with an intensity a fraction (typically 10%) of the peak. The process is repeated on the new image, where the peak may be the same as before, or a new peak may have emerged. The process continues until only *noise* is left, and the corrected image is then built up from the record of the subtractions of the PSF. See also: *maximum entropy method.*

cleaning

The surfaces of lenses and mirrors may acquire a layer of dust or become dirty in other ways over a period of time. However, cleaning them should only be undertaken following the directions from the manufacturer or the coatings on the optical surfaces may be damaged or even the glass scratched. Some proprietary cleaning methods may be found in advertisements in the popular astronomy magazines. Large telescope mirrors are often cleaned nowadays by drifting carbon dioxide snow across them. Never be tempted to try and clean the optics of your telescope using household cleaning materials.

cloud chamber

A detector of high-energy charged particles, such as secondary *cosmic rays*. An enclosure is super-saturated with an appropriate vapour. The passage of a particle through the enclosure initiates the formation of small droplets along its track (i.e. a cloud), which can then be seen directly.

C

CNES

See *Centre National d'Etudes Spatiales.*

co-altitude

A synonym for *zenith distance.*

COAST

See *Cambridge optical aperture synthesis telescope.*

coating

See *blooming* and *mirror coating.*

coded array mask

A technique used at *X-ray* wavelengths to obtain images of extended sources. For lower energy X-rays, grazing incidence telescopes now supersede the technique, but it is still used for high energy X-rays. The technique is essentially that of the *pinhole camera*, but with many pinholes (actually square apertures) set out in a known distribution with opaque segments between them, to form the mask. A position-sensitive *X-ray detector* is placed behind the mask, and the instrument scanned across the source. Fourier analysis of the output then retrieves the image. See also: *Hadamard mask.*

coelostat

A device for directing the light from any part of the sky into any fixed direction, and without any rotation of the final image. It is used to feed large, cumbersome or delicate instruments such as *solar telescopes*. It usually comprises two plane mirrors both on driven mountings. A sideriostat uses just a single plane mirror on an *equatorial mounting* to feed a telescope mounted parallel to the Earth's rotation axis. In this latter system, however, the image rotates as the object moves across the sky. See also: *Coudé focus; Nasmyth focus; heliostat.*

coherence

The degree to which two or more waves are correlated (in step) with each other. The radiation from *lasers* and *masers* is an example of 100% coherence. See also: *autocorrelation.*

coherence bandwidth

The range of *frequencies* over which two or more waves are *coherent.*

coherence interval

The time for which two or more waves remain *coherent.*

coherence length

The distance over which two or more waves are *coherent.*

coherent

See *coherence.*

coincidence detector

A device to allow the direction an incoming particle or photon to be deduced. It is usually used for *gamma rays* or *cosmic rays* where direct imaging is not possible. The device requires two detectors of the radiation that are separated by a short distance. Overall detection only occurs when both detectors are triggered simultaneously. The photon or particle must then have travelled down the line joining the two individual detectors, and so the direction in space of its source may be deduced.

co-latitude

See *celestial latitude.*

cold camera

See *cooling.*

collimation

The process of aligning correctly the optical components of an instrument. Within astronomy it usually refers to the alignment of mirrors and lenses within a telescope. For most telescope designs, the instrument is collimated when, on looking through the eyepiece holder without an eyepiece in place, all mirrors, lenses, their images and reflections are concentric. See also: *laser collimator.*

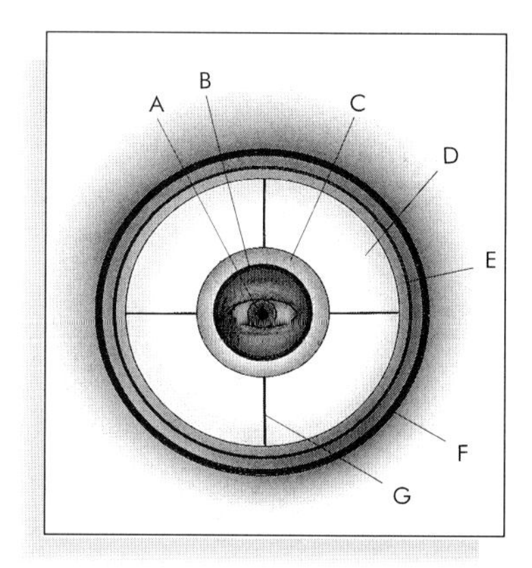

The view through the eyepiece holder (without the eyepiece present) for a correctly collimated Newtonian telescope.
A Eye reflected in the primary and the secondary (twice).
B Eyepiece holder reflected in the primary and the secondary (twice).
C Secondary reflected in the primary and the secondary.
D Primary reflected in the secondary.
E Secondary seen directly.
F Eyepiece holder seen directly.
G Spider reflected in the primary and the secondary.

collimation error

A mis-alignment between the *optical axis* of a telescope and the *declination axis* of its *equatorial mounting*. The two should be exactly orthogonal if the telescope is to *track* objects correctly.

collimator

1 A lens or mirror used to produce a parallel beam of light as its output. It is most frequently encountered as a part of a spectroscope.
2 At *X-ray* and *gamma ray* wavelengths it is a device to restrict the field of view so that an image of an extended object may be built up by scanning across it. At its simplest it consists of an array of rectangular tubes placed before the detector and is then known, slightly inaccurately, as a honeycomb collimator. See also: *modulation collimator.*

Collinder catalog

A catalog of 471 open clusters, published by Lund observatory in 1931.

collinear array

See *array.*

color–color diagram

A scatter plot using two different *color indices*. For mid-temperature stars, the U–B/B–V color-color diagram allows main sequence and giant stars to be separated.

color balance

In a color image, the balance, or relative intensity, of the monochromatic images from which it is produced. For a normal image, the aim of the color balance will usually be to reproduce the appearance of the object as seen directly by the eye, or as the eye would see it if our color vision were sensitive enough. Most photographic color emulsions are balanced to be correct for short exposures and become incorrect at exposures of more than a few seconds. Long exposure color images may then sometimes have their color balance corrected by *cooling* the emulsion. For some purposes the color balance may be deliberated skewed in order to enhance some feature of the image. In *false color* images, especially where one or more of the monochromatic images has been obtained at a wavelength outside the visible region, there is no correct color balance, and so the color balance that shows the features of interest at their best will normally be used. See also: *image processing.*

C

color coding

See *false color image.*

color excess

The difference between the observed *color index* of an object and its value in the absence of *interstellar absorption* (known as the intrinsic color index), symbolized by E_{B-V}, etc. Interstellar absorption increases as the wavelength decreases, so the color excess is always positive. See also: *reddening ratio.*

color film

Photographic film with three types of photographic emulsion sensitive to different parts of the spectrum that produces color images directly. Although it can be used for astronomical imaging, the *color balance* will change on long exposures due to different rates of *reciprocity failure* in the three emulsions. Color images are therefore often better obtained by taking three separate exposures on black and white film through different filters and then combining these to produce the color image.

color index

The difference between the *magnitudes* of an object at two different *wavelengths.* The most frequently encountered color indices are for the *UBV photometric system.* The (U–B) and (B–V) color indices are then the differences between the magnitudes of the object seen through the U and B, and the B and V filters, respectively. Color indices may be used to obtain details of an object through the relatively simple observational technique of *photometry.* Thus the *color temperature* of a star, T, and its B–V color index are related by:

$$T = \frac{8540}{(B-V)+0.865} K$$

for temperatures up to about 10,000 K.

color-magnitude diagram

A version of the Hertzsprung–Russell diagram with a *color index* (usually B–V) replacing the *spectral type.*

color temperature

The temperature of a black body that has the same *color index* as the observed object.

colure

Meridians of constant *right ascension.* The solsticial colure is a great circle defined by RA = 6 h and 18 h, the equinoctial colure, the great circle defined by RA = 0 h and 12 h.

C

coma

A fault of images in which the off-axis images consist of a series of circles of increasing size, shifted progressively towards or further from the optical axis for annular zones of the optics of increasing sizes. The result for point sources is a fan shaped image somewhat resembling a comet and from which the name derives (the term coma is also used as a synonym for the head of a comet). See also: *aberration (optical).*

The effect of coma on the image of a point source.

combined magnitude

The *magnitude* of two objects so close together that they are seen as a single object (for example close double stars). For two objects, A and B, of magnitudes, m_A and m_B, the combined magnitude, m_C, is given by:

$$m_C = m_A - 2.5 \log_{10} (1 + 10^{0.4(m_A - m_B)}).$$

comes

See *double star.*

comet filter

A *filter* that transmits light from about 460 nm to 550 nm so that the emission lines of molecular carbon (known as the Swan bands) are transmitted. Such a filter will help to enhance the views of the ion tail of a comet. See also: *light pollution rejection filter; nebula filter.*

comet nomenclature

On discovery, comets are designated by the year of discovery and a running letter plus, usually, the discoverer's name (e.g. comet Kohoutek 1973f). Once an orbit has been determined, then the name changes to the year of perihelion passage plus a running Roman numeral giving its order of perihelion passage. Comets with periods under 200 years are additionally denoted by 'P' to denote a short period and often abbreviated to just that plus a name (e.g. comet West 1976 VI, comet P/Halley). New discoveries of comets must be reported to the Minor Planet Center via the *Central Bureau for Astronomical Telegrams* in order to be recognized.

comparator

See *blink comparator.*

comparison spectrum

A spectrum, usually of *emission lines* from an artificial source, that is imaged alongside the *spectrum* of an astronomical object in order to enable the *wavelengths* of lines within the main spectrum to be determined.

comparison star

A *standard star* that is compared with the star being observed in order to determine the latter's properties.

compass points

1 The points on the *horizon* that are directly north, east, south and west of the observer, also known as the cardinal points.
2 The points on the horizon that are indicated by a magnetic compass as north, east, south and west, and therefore differ from the true directions as a result of the *compass variation.*

compass variation
The difference between the direction shown as north by a magnetic compass and the true direction of north. Since the magnetic and geographic poles do not coincide (the magnetic pole is currently in northern Canada), and the magnetic pole moves, the compass variation changes with time and with the observer's position on the Earth. Its value can usually be found from large-scale maps.

compound lens
A *lens* made from two or more individual simple lenses, such as an *achromat* or *apochromat.*

compression
See *image compression.*

Compton interaction detector
A *gamma ray* detector that operates by detecting the high-energy electrons produced by gamma rays through Compton scattering. The electrons are detected in *scintillation detectors.* Usually two or more detectors separated by a meter or so are used in order to determine the direction of the gamma ray.

computer modeling
See *modeling.*

concentration
The actual diameter of the central disk of the image of a point source produced by a telescope. The minimum size is that of the *Airy disk*, but it is usually larger than this through the effects of *aberrations*, and atmospheric *seeing*. It is also used in the sense of the percentage of the total light from the star that the telescope concentrates into the central disk of the image, for which values of 60% to 70% are typical for modern large telescopes. See also: *Strehl ratio.*

condensation
Water precipitating out of warm air onto a colder surface, it can cause rapid deterioration of surface coatings on optical components and of course, rusting of any steel parts of the mounting. Condensation can be a problem with portable instruments if they are brought back into a warm house after use, and precautions should be taken to ensure that they dry off as quickly as possible (but do not wipe the optical surfaces dry!). For instruments permanently mounted in an observatory, condensation can be a problem through diurnal temperature changes. The use of fans to keep air circulating or de-humidifiers may be required. See also: *dew cap.*

cone cell
See *retina.*

confocal
An arrangement for two optical components such that they are separated by the sum of their focal lengths. This is the normal setting for a telescope used visually. See also: *afocal.*

confusion limit
A limit to the faintest objects discernable on an image that is imposed by numerous fainter objects blurring together. For example surveys at radio wavelengths by low angular resolution instruments or observations of the centers of galaxies and globular clusters are likely to be confusion limited.

conic section

A geometrical shape ranging from the circle, through the ellipse and parabola to the hyperbola. Most optical surfaces have cross sections that are conic sections (circular for most *lenses*, parabolic and hyperbolic for *primary mirrors*, hyperbolic and elliptical for *secondary mirrors*). The name derives from the shapes of slices taken at various angles through a cone (circular for a slice orthogonal to the cone's axis, parabolic for a slice parallel to one side of the cone, etc.). See also: *eccentricity.*

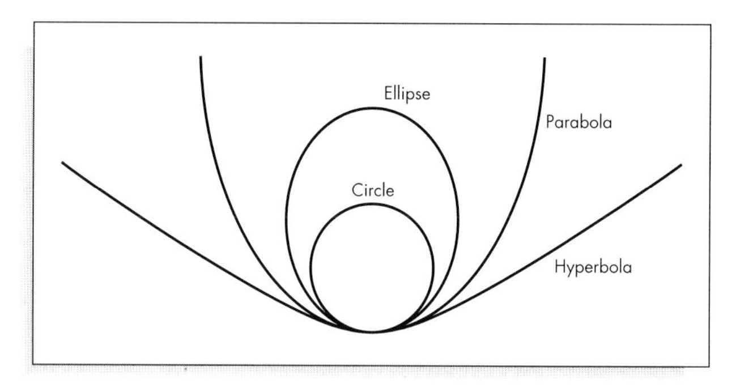

The shapes of the conic sections – circle, ellipse, parabola and hyperbola.

conjunction

The instant when two moving objects, such as planets, a planet and the Moon, or a planet and the Sun have the same *celestial longitude.* Often used colloquially as a synonym for an *appulse.* See also: *inferior conjunction; superior conjunction; opposition; aspect.*

constellation

See panel (pages 49–50).

constringence

A measure of the *dispersion*, or the way in which the *refractive index* of an optical material varies with wavelength, symbol, ν. It is given by:

$$\nu = \frac{\mu_{589} - 1}{\mu_{486} - \mu_{656}},$$

where m_{486} is the refractive index at 486 nm (Balmer H-b line), m_{589} is the refractive index at 589 nm (sodium D line) and m_{656} is the refractive index at 656 nm (Balmer H-α line), The constringence has values of 30 to 35 for *flint glass*, and 55 to 60 for *crown glass.* Note that the higher the numerical value of the constringence, the less the refractive indices at different wavelengths differ from one another. See also: *dispersive power.*

constructive interference

See *interference.*

continuous spectrum

A *spectrum* with a continuous series of uninterrupted colors (i.e. no *emission* or *absorption lines*), usually extending from red through to violet, but also occurring at all other wavelength ranges. A continuous spectrum is obtained from hot solids or liquids or high-pressure gases.

continuum

The portions of a *spectrum* between *absorption* or *emission lines* (i.e. featureless segments of a spectrum). The intensities, or *equivalent widths* of spectrum lines are measured relative to the intensity of the nearby continuum. In the absence of spectrum lines, the continuum becomes the *continuous spectrum.*

contour image

An image in which the various levels of intensity are represented by lines.

constellation

One of 88 areas, dividing up the whole sky (see Appendix for full list). Many constellations, especially those associated with the zodiac, date back to pre-history. Their names are traditional and the shapes outlined by the principal stars rarely bear any relationship to the object or person that they supposedly represent. Southern constellations were mostly named at the end of the sixteenth century. A few minor constellations were added in the 1750s. The main use today of constellations is to provide a quick means of finding your way around the sky. For this purpose it is usually sufficient to be able to recognize the 15 or 20 most prominent constellations visible from your latitude. Constellations are also used as a basis for stellar nomenclature such as the *Bayer system* (e.g. β Cyg), the *Flamsteed system* (e.g. 61 Cyg) and *variable star nomenclature* (e.g. SU Cyg, V1057 Cyg).

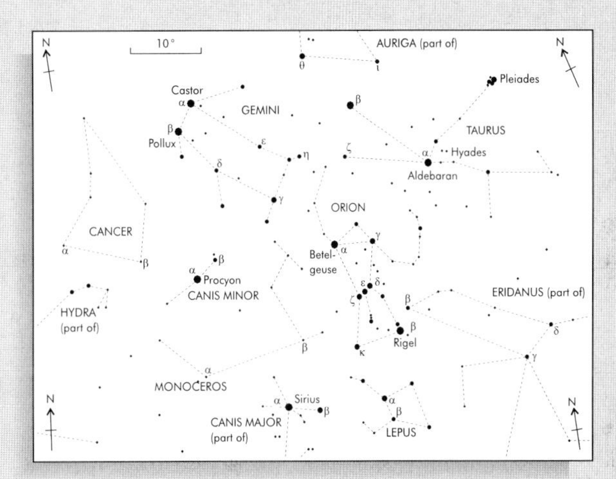

Orion and the nearby constellations.

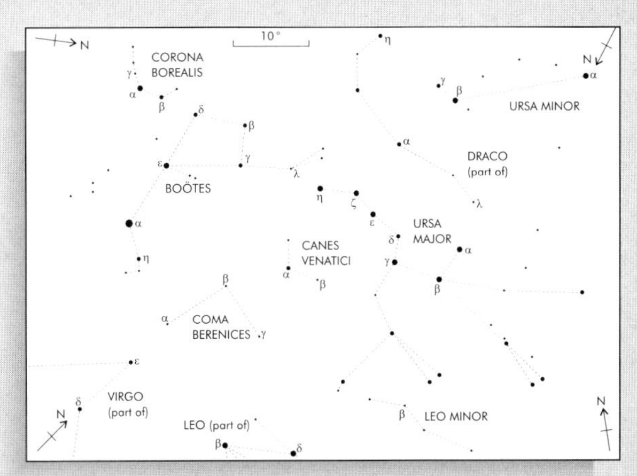

Ursa Major, Polaris and the nearby constellations.

C

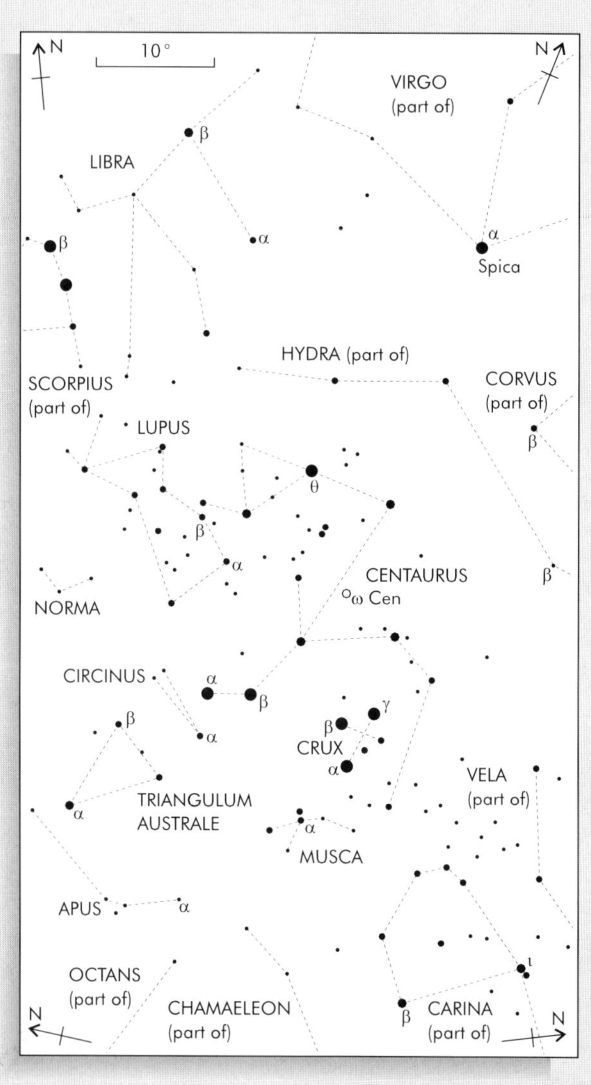

Crux and the nearby constellations.

contrast
The difference between the light and dark portions of an object or an image.

contrast index
A measure of the contrast of a *photographic emulsion*. It is obtained from the toe and slope of the linear portion of the *characteristic curve* up to an *optical density* of 2.0. See also: *gamma*.

C

contrast stretching
A synonym for *grey scaling*.

convergent point
The point in the sky towards which the *proper motions* of the stars in a galactic cluster appear to converge or from which they appear to diverge. It represents the direction in space of the relative velocity between the solar system and the star cluster. It is used in the *moving cluster method* for determining the distances of stars. See also: *radiant*.

converging lens
A *lens* that causes an incident parallel beam of light to converge to a *real image* at its *focus*. Also known as a positive lens. See also: *diverging lens*.

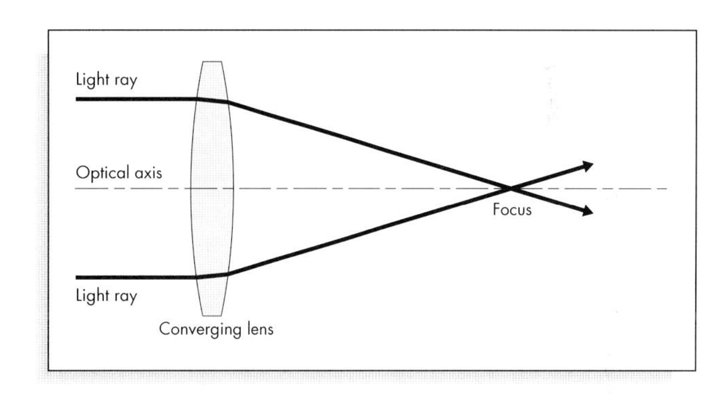

Light paths through a converging lens.

convolution
A mathematical operation describing the blurring effect of an instrument. The data prior to detection (known as the true data), T, is smoothed by the *point spread function*, I, of the instrument to produce the observed data, O:

$$O(x) = \int_{-\infty}^{\infty} T(y)I(x-y)dy .$$

See also: *deconvolution*.

cooling
A technique applied to *detectors* in order to reduce the thermal *noise* generated within them. The detector, and sometimes its immediately associated electronics, is/are operated at a temperature below ambient. The cooling is usually effected by the use of a Peltier device, dry ice, liquid nitrogen or helium. Small *CCD* detectors are usually cooled by Peltier coolers to 40 or 50 degrees below ambient. Larger CCDs and many other detectors, especially those operating in the infrared, are cooled much more. In some cases liquid helium under reduced pressure is used to give an operating temperature less than 3 K. Photographic emulsion may also be cooled (usually with an acetone/dry ice mixture). The purpose of this, however, is to reduce *reciprocity failure* or to improve *color balance*.

coordinated universal time
The basis of *civil time*, abbreviated UTC (from universel temps coordonné). UTC is derived from *international atomic time* by the insertion of leap seconds occasionally in order to keep it within 0.9 s of *universal time*. (Tabulated annually in the Astronomical *Almanac*.)

Copenhagen University observatory
An optical observatory based in Copenhagen, Denmark, with telescopes at the *European Southern Observatory.*

Córdoba Durchmusterung catalog
See *Bonner Durchmusterung catalog.*

Cornell Atacama telescope
A proposed infrared/sub-millimeter telescope with a 15 m segmented primary mirror to be based in the Atacama desert. See homepage at http://www.astro.cornell.edu/atacama/atacama.html.

Cornu prism
A *prism* made from crystalline quartz for use in the *ultraviolet.* Since quartz is *birefringent,* the optical axis has to be parallel to the base of the prism. Quartz is also optically active (i.e. it causes the plane of *polarization* to rotate), and so the prism has to be made in two halves from left and right-handed crystals that are then cemented together.

Coronado
A commercial firm manufacturing H-α filters.

coronagraph
1 Solar – An instrument for observing the solar corona. The coronagraph produces an artificial eclipse by obscuring the solar photospheric disk with an occulting disk. The occulting disk may be inside the instrument, at the focus of the objective, or placed some distance in front of the objective. Since the corona is at most a millionth of the intensity of the photosphere, great precautions need to be taken to eliminate any scattered light. On the Earth, this means using a simple lens of very high purity as the objective, keeping all optical surfaces scrupulously clean, using a *narrow band filter* centered on a coronal *emission line* and siting the instrument at high altitude. Such instruments can then observe the solar corona out to one or two solar radii. Spacecraft-borne coronagraphs can detect the corona out to five or ten solar radii. (CAUTION – always take appropriate precautions when observing the Sun or damage to the eye and/or telescope can result.)
2 Stellar – An instrument for observing a faint object close to a bright object. The design aims to produce a very clean *point spread function* that drops of rapidly at its edge and is without diffraction spikes or fringes. The bright object is then obscured with an occulting disk, as in the solar coronagraph. Instruments are currently being considered to allow the direct observation of planets next to stars.

correcting lens
A lens specifically designed to correct *aberrations* produced by the remainder of the optical system. The lenses are usually thin and of low or zero overall power. The use of *aspheric surfaces* and exotic optical materials is quite common. Correcting lenses are found as an integral part of the design of *Schmidt cameras, Schmidt-Cassegrain* and *Maksutov telescopes.* Correcting lenses are also often used before the *prime focuses* of large telescopes where they provide sharply focused images over a relatively wide field of view (for example the two degree field (2df) on the *AAT*). See also: *catadioptric telescopes.*

correction curve
A plot of the known errors of an instrument that is used to correct the readings from that instrument. See also: *characteristic curve; calibration.*

correlation coefficient

A statistical test of the degree to which pairs of data measurements are correlated with each other (i.e. how close the data is to a straight-line graph).

C

correlation detection

A technique for improving detection in low *signal/noise* situations and when the form of the signal is known. For example when detecting planets using Earth-based *radar*, the expected time of return of the pulse and its shape will be known approximately from previous work, and the new data is found in the exact return time and pulse shape.

correlation receiver

A radio receiver in which the signals from two *antennas* are multiplied together. Random noise then averages towards zero, while the real signal gives a steady output. Over a period of time the signal output will show a *fringe pattern* as the object moves across the sky similar to that from a *phase-switched interferometer*, except that the signal from the latter is symmetrical about zero.

correlator

A device used as a part of the processing of the outputs from the *antennas* of a radio *interferometer* to multiply the two signals together. See also: *autocorrelator.*

cosmic ray detector

A detector for *cosmic rays.*

Since the Earth's atmosphere is opaque to primary cosmic rays, detectors for these have to be flown on rockets or spacecraft. They can also normally be used to detect X-ray and gamma rays and include; *proportional counters*, *scintillation detectors*, *nuclear emulsions* and *spark detectors.*

Detectors for secondary cosmic rays include all of those for primary cosmic rays plus cloud chambers and detectors of the *čerenkov* radiation produced by the particles high in the Earth's atmosphere. Secondary cosmic ray detectors are usually used in groups in order to detect different components of the shower of particles and photons efficiently. Several such groups are then laid out in an array covering a square kilometer of more in order that a representative sample of the shower is captured.

cosmic rays

1 Primary cosmic rays are very high-energy charged particles permeating most of space and probably originating from supernovas and active galaxies.

2 Secondary cosmic rays are high-energy sub-atomic particles of many varieties, plus gamma rays, which result from the interaction of primary cosmic rays with atomic nuclei high in the Earth's atmosphere. For the higher energy primary cosmic rays, a shower of millions of particles and photons results which may survive down to the Earth's surface. Secondary *cosmic ray spikes* are an important *noise* source in a *CCD detector*, since they generate numerous electron-hole pairs on passing through the device.

cosmic ray spike

One or more *pixels* of a *CCD* image that are brighter than they should be due to ionizations arising from the passage of a *cosmic ray* through the CCD during the exposure. They are usually removed during *data processing* by replacing the affected pixel intensity with the average intensity of the surrounding pixels, although this does not restore the lost information.

C

Coudé focus

A focus position for *equatorially mounted* telescopes that is fixed in space irrespective of the telescope orientation. There are a number of variations on the design, but the basic system has a plane mirror that reflects the light beam from the telescope down the hollow *declination axis* and a second plane mirror to reflect it through the hollow *polar axis*. The fixed focus thus emerges from the end of the polar axis. Most large equatorially mounted telescopes have a Coudé focus because it means that bulky, unwieldy, flimsy or delicate equipment can be used since it does not have to move with the telescope. The Coudé focus has the disadvantages of a small *field of view* that also rotates as the telescope *tracks* across the sky. An *image de-rotator* is therefore needed to provide stationary images. See also: *coelostat; Nasmyth focus; Springfield mounting.*

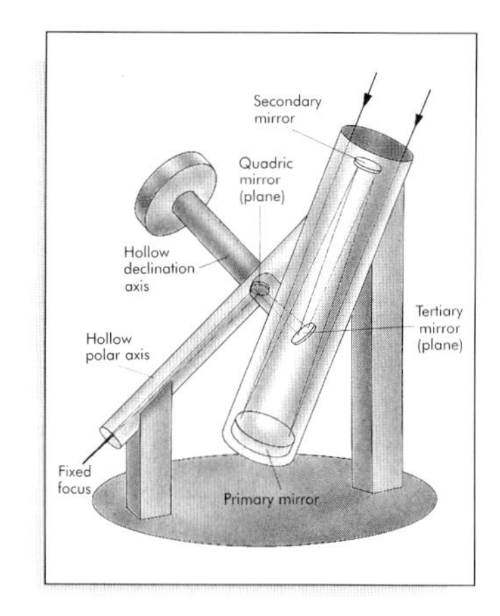

Light paths within a Coudé telescope system.

Couder telescope

A variant of the *Gregorian telescope*, which uses aspheric mirrors to provide a highly corrected image. The field of view, however, is curved and located inside the instrument. The design is rarely encountered since it has no advantages over the *Schmidt camera*.

counter weight

A weight added to a telescope mounting so that the telescope tube or the mounting as a whole is balanced. See also: *German mounting; modified English mounting.*

crater counting

A technique for dating areas of the surface of the Moon and other atmosphere-less planets and satellites. The number of craters of a given size, per unit area, is determined. The time since that area was last cleared by geological activity or by a large impact may then be estimated from the rate at which such craters are produced. The technique is the only way of finding absolute ages for surfaces where actual rock samples are not available. However, it is very imprecise because the cratering rate and the way that has changed with time is poorly known.

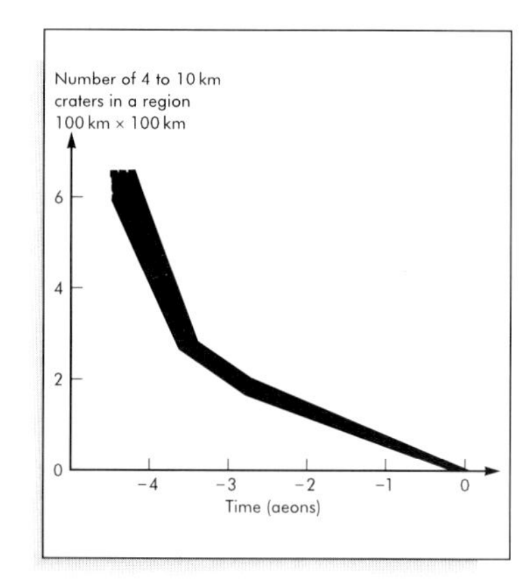

The variation of crater density with time, used for dating planetary surfaces by crater counting.

crescent phase

The shape of the Moon, Mercury or Venus, when less than half of the illuminated surface is visible. See also: *gibbous phase.*

Crimean astrophysical observatory

An optical and radio observatory at Nauchny in the Ukraine, altitude 600 m, and with several out-stations. It houses a 2.6m and several other optical telescopes and a 22m radio telescope.

critical frequency

A synonym for *Nyquist frequency*.

cross axis mounting

See *modified English mounting.*

cross correlation spectroscope

A *spectroscope* designed to measure *radial velocities* quickly and accurately. A mask is placed over the spectrum produced by the spectroscope, and the total energy passing through measured. The mask is a negative version of the spectrum. The mask is moved along the spectrum, and when it coincides exactly with the spectrum, there is a sharp drop in the output. The position of the mask when the drop occurs can be calibrated to read the radial velocity directly. The technique is especially useful for binary stars since the only change in the spectrum from one night to another is the amount of the *Doppler shift*.

cross disperser

A low power *spectroscope*, often using a *prism* that is used to separate the over lapping high order spectra produced by *echelle gratings* and *Fabry–Perot etalons*. The dispersion of the cross disperser is orthogonal to that of the principal spectroscope and so an array of short segments of the whole spectrum is produced which may then be joined together to give the entire spectrum.

cross talk

A synonym for *blooming (2)*.

cross wire eyepiece

An eyepiece that has two fine orthogonal threads mounted at its internal focal plane. The threads can usually be focused separately from focusing on the image, so that both can be seen sharply. Often the threads can be illuminated so that they may be seen against a black background. The cross-wire eyepiece is used on a *finder telescope* for setting the main telescope onto the desired object, and on a *guide telescope* for keeping an object accurately centered throughout a long exposure. Thread from spider webs was commonly used for the cross wires at one time, but it has now largely been replaced by thread obtained by unravelling dental floss. See also: *bi-filar micrometer; graticule.*

crown glass

A type of glass with a *refractive index* around 1.5 and a *constringence* of 55 to 60 that is often used to form the converging component of an *achromatic lens*. See also: *flint glass.*

cryogenic

An instrument or process operating at low temperatures.

cryostat

A device for *cooling* a detector. It often takes the form of a vacuum flask, or Dewar, which contains a liquefied gas such as nitrogen or helium, and it may also contain some of the optical and electronic components.

CTIO

See *Cerro Tololo Inter-American observatory*.

culmination

The instant when an object crosses the observer's *prime meridian*. Non-*circumpolar* objects culminate just once a day. They are then south or north of the observer, with an *hour angle* of 0 h and at their highest point in the sky. Circumpolar objects cross the prime meridian twice; giving an upper culmination when they are highest in the sky (hour angle 0 h), and a lower culmination when they are at their lowest (hour angle 18 h). See also: *transit.*

curvature of field

See *field curvature.*

curve of growth

A plot of the way in which a *spectrum line's* strength (*equivalent width*) increases with increasing numbers of the atoms, ions or molecules producing it. It is the basis of a method of determining the abundances of elements in stars, but which has now largely been replaced for the purpose by computer *modeling* of stellar atmospheres.

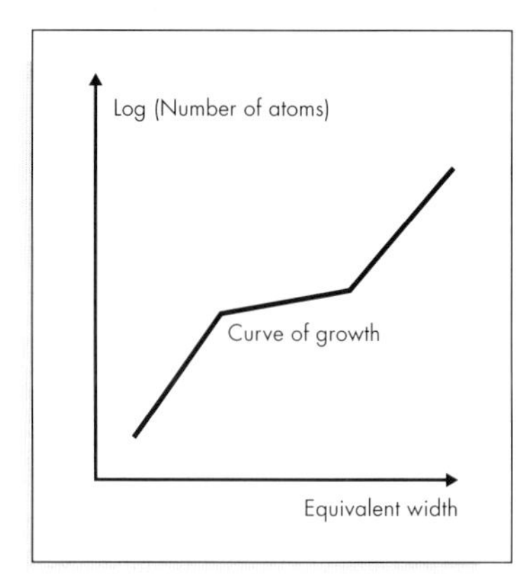

A typical curve of growth.

cut-off filter

A *filter* that blocks radiation above or below a certain wavelength, whilst allowing it through it on the other side of that limit. See also: *H-α cut-off filter; violet cut-off filter.*

cut-off wavelength

The upper or lower limit of the wavelength range of a filter, detector, etc.

D*

See *normalized detectivity.*

Dall-Kirkham telescope

A variant of the *Cassegrain telescope* using an elliptical *primary mirror* and a spheroidal *secondary mirror.* It has good images for a small *field of view* around the *optical axis* but the quality deteriorates rapidly away from that region due to severe *coma.* It is best suited to visual planetary and similar observations.

damping

In a mechanical system, a degree of deliberately introduced friction that reduces unwanted vibration. In an electronic system, a combination of resistances, capacitances and inductances that similarly reduces unwanted electrical oscillations.

Danjon astrolabe

An instrument designed to determine stellar positions and local time, also known as the impersonal astrolabe, since it is unaffected by focusing errors. It allows observation of stars at an altitude of 60°. Two separate beams of light from the star are fed into the instrument via a 60° prism, one of them having been reflected from a bath of mercury in order to give a precise zenith direction. The beams are split by a *Wollaston prism,* and the two beams parallel to the optical axis observed. Measurement of the star's position is made by moving the prism along the optical axis until the images merge, with diurnal motion being compensated by movement of the Wollaston prism. See also: *transit telescope; photographic zenith tube.*

Danjon scale

A five-point scale for gauging the depth of a lunar total eclipse:

0 Very dark eclipse. Moon almost invisible.
1 Dark Eclipse, grey or brown in color. Surface details seen only with difficulty.
2 Deep red eclipse. Very dark central shadow, with the outer edge of umbra relatively bright.
3 Brick-red eclipse. The umbral shadow usually has a bright edge.
4 Bright red or orange eclipse. The umbral shadow has a very bright edge.

dark adaptation

The phenomenon whereby the eye's sensitivity improves dramatically after a few minutes in the dark, also known as night vision. A small component of dark adaptation is the increase in the size of the pupil of the eye. Under bright conditions this is typically 2 to 3 mm in diameter, increasing to 7 mm under low light conditions and thus allowing 7 or 8 times as much light into the eye.

The main improvement, however, is due to the increasing sensitivity of the rod cells in the retina that provide monochromatic vision. Monochromatic vision is due to absorption of light by a molecule called rhodopsin or visual purple. This is present in the rod cells. Color vision is due to cone cells in the retina and there are three types responding to blue, green and yellow-red respectively. Under bright light conditions, the rhodopsin in the rod cells is exhausted, and vision is in color via the cone cells. In the dark, the rhodopsin regenerates over a period of about 20 minutes. When the rhodopsin is fully regenerated, the rod cells are about 100 times more sensitive than the cone cells, and the rod cells then provide vision, but colors can no longer be distinguished.

A brief exposure to bright light will again exhaust the rhodopsin, so observers usually try to work under low light conditions, or in the dark, in order to preserve their night vision. The longest wavelength to which rhodopsin is sensitive, however, is about 600 nm, while the yellow-red cone cells' sensitivity extends to 700 nm. It is possible therefore to use a very deep red illumination without affecting night vision. However, most red lights used for this purpose, including red LEDs, are too orange for this to work well. See also: *averted vision; retina.*

dark current

See *noise.*

dark exposure

The subtraction of a blank image of the same exposure length as the main image, known as the dark exposure, to reduce the background noise in CCD images.

dark noise

See *noise.*

data analysis

Data analysis is the conversion of the data after *data reduction* into an astronomically interesting form.

At its simplest it may be just the comparison of a new measurement of the magnitude of a star with one made previously to see if it has varied in brightness. At its most complex it may involve the detailed computer modeling of a stellar or planetary atmosphere or interior, interstellar nebula, galaxy, cluster, etc. in order to determine rates of nucleosynthesis, element abundances, velocity, temperature or density structures, sources of energy and so on, it may need to include data of many different types, from many different sources, require complex theories, abstruse mathematics, and advanced concepts from physics and other sciences.

The processes involved in data analysis thus depend upon the nature of the observations and the purpose for which they are intended.

data processing

A general name for numerous techniques applied to *raw data* from a telescope, etc. to convert it into an astronomically useful and interesting form. It sub-divides into *data analysis* and *data reduction*. See also: *image processing.*

data reduction

The process of correcting measurements for known errors or problems and/or converting it to a standard format for further processing. It is "mechanical" side of *data processing*. Parts or all of data reduction can often be done automatically if the data is stored in the computer or is available in computer-readable form.

The exact stages involved in data processing vary with the observations and with the ultimate purpose for which they are to be used. But some examples will demonstrate the type of operation that is involved:

- removal of *cosmic ray spikes* from *CCD* images
- removal of the background *noise*, and the *flat fielding* of CCD images
- *grey scaling*, false color representation, and other *image processing* techniques
- determination of the response (*characteristic curve*) of a *photographic emulsion* and the conversion of the photographic density of the original image back into intensity
- correction of geometrical distortions in the image
- *calibration*
- *smoothing* or other noise reduction procedures such as adding together several images

- correction for expansion or contraction or other temperature-induced defects
- reduction of the blurring effect (see *point spread function*) of the telescope and other ancillary instruments used to obtain the data
- removal of electrical mains supply interference or of other cyclic defects
- conversion of intensity values into stellar magnitudes
- calibration of the wavelength along a *spectrum*
- correction for *atmospheric extinction*
- correction for the effect of Earth's velocity on wavelengths (*Doppler shift*) and position in the sky (*aberration*)
- correction for the effects of any *filters* used while obtaining the observations
- calculation of average values, *standard deviations* and *standard errors of the mean* for a set of measurements.

D

David Dunlap observatory
An optical observatory at Richmond Hill, Ontario.

Dawes limit
An empirical measure of the resolution of a telescope, given by 4.56″/D, where D is the *objective* diameter in inches (equivalent to 0.116″/D for D in meters). See also: *Rayleigh resolution.*

daylight saving time
See *civil time.*

day number*
See *Besselian day number* or *independent day number* or *Julian date.*

Daystar
A commercial firm manufacturing H-α filters.

dead time
A brief time interval occurring after some types of detector have detected one event, during which they have reduced or zero sensitivity for subsequent events.

Dec
See *declination.*

declination
See panel (page 60).

declination axis
On an *equatorial telescope mounting*, the axis, movement around which changes the telescope's *declination*. The declination axis is orthogonal to the *polar axis.*

declination circle
See *setting circle.*

deconvolution
A mathematical process for removing the blurring effect of an instrument. Most modern approaches are based upon *Fourier transforms*. The Fourier transform of the unblurred or true data, F(T), is given by the Fourier transform of the observed data,

declination and right ascension

A coordinate system, for giving the positions of objects in the sky. The declination (symbol, Dec or δ) is the angle, in degrees, up or down from the equator, going from 0° to ±90°. See also: *celestial sphere.*

The right ascension (symbol, RA or α) is the angle, in *hours; minutes and seconds*, around the *ecliptic* from the *First Point of Aries*, going from 0h to 24h, and measured eastwards. (RA and Dec are tabulated annually in the Astronomical *Almanac* for the Sun, Moon, planets and some asteroids.) See also: *hour angle; sidereal time; celestial sphere.*

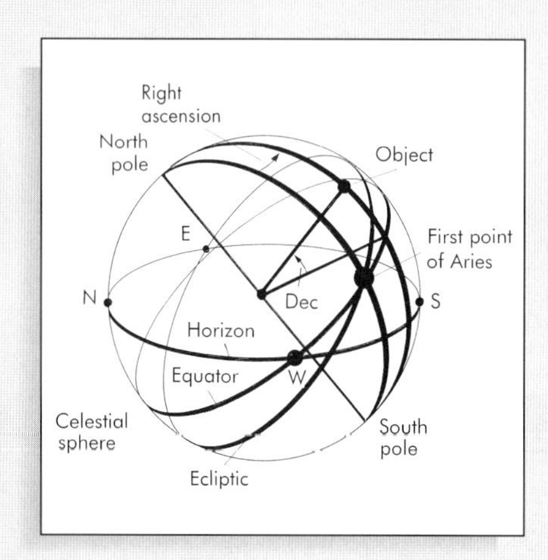

The definition of right ascension and declination.

The declination setting circle on a small telescope.

The right ascension and hour angle setting circles on a small telescope.

F(O), divided by the Fourier transform of the *point spread function*, F(I). The unblurred data is then obtained by taking the inverse transform, F^{-1}:

$$T = F^{-1}\left[\frac{F(O)}{F(I)}\right].$$

See also: *CLEAN; convolution; maximum entropy method.*

deep sky filter
See *light pollution rejection filter.*

deep sky object
Any object beyond the solar system, but especially those outside our own galaxy.

deep space network
A network of large radio telescopes distributed around the world to allow continuous communication with spacecraft.

definition
A term for the observed angular *resolution* and image clarity of a telescope, especially when observing *extended objects.*

degree
A measure of angle defined as 1/360 of a complete circle, symbol, °. It is sub-divided into *arc minutes* and *arc seconds.* $1° = 60' = 3600''$. See also: *radian.*

delay line
A length of cable added to the transmission line between a radio *antenna* and its *receiver* in order to produce a delay in the arrival time of the signal. Delay lines are usually used in order to change the phase relationship between signals from two or more antennas.

density
See *optical density.*

depolarizer
A device that converts *polarized radiation* to unpolarized radiation. Depolarizers usually operate by averaging the polarization towards zero over time, space or wavelength. They are used in some *polarimeters*, and with detectors that may have an intrinsic sensitivity to polarization, such as *photomultipliers.*

depth of focus
The range of distances for an object being observed or imaged through an optical system over which it remains in acceptable focus. On camera lenses the *f-ratio* is variable in order to allow the depth of focus to be changed. The smaller the f-ratio (i.e. the larger the aperture), the smaller the depth of focus.

descending node
See *node.*

destructive interference
See *interference.*

D

detective quantum efficiency

A measure of the *sensitivity* of a detector. It is given by the square of the ratio of the *signal to noise ratios* of the output from the detector to the input.

detectivity

A measure of the *sensitivity* of a detector. It is the reciprocal of *noise equivalent power*. See also: *normalized detectivity.*

detector

Any device which converts incoming radiation, particles, etc. into a detectable signal. See also: *CCD detector; photographic emulsion; pin photodiode; bolometer; radio receiver; X-ray detector; neutrino detector; gravity wave detector.*

developing

A chemical process that converts the *latent image* in a *photographic emulsion* to a real image by converting exposed silver bromide crystals to silver particles. See also: *fixing.*

Dewar

See *cryostat*.

dew cap

An extension to the *telescope tube* of a *refractor* or a *catadioptric telescope* beyond the *objective* or *correcting lens* by typically 1.5 to 3 times the tube diameter. The dew cap inhibits the formation of dew on the objective. Low power, low voltage heaters may be incorporated into the dew cap to increase its effectiveness. Such heaters will of course cause the seeing to deteriorate to some extent, but they are only likely to be used when condensation is so bad that observing would otherwise be impossible anyway.

A dew cap. (Image reproduced by courtesy of Meade Instrument Corp.)

dew remover

A device for removing dew from misted-up optical surfaces. Some *dew caps* incorporate heaters for this purpose. Alternatively a small hair drier produced for campers, etc. and therefore operating from a battery, not the mains power supply, may be used to blow warm air over the optical surfaces affected. Be careful not to over-heat the surfaces.

DO NOT be tempted to wipe the surfaces dry. At the very least this will leave them smeary, and may damage coatings on the optics or even scratch the glass.

dew shield

a synonym for *dew cap.*

diagonal

A plane mirror or sometimes a 90° *prism* that is placed at 45° to the optical axis and reflects the light through 90°. Diagonals form the *secondary mirrors* in *Newtonian telescopes*, and the tertiary and quarternary mirrors in *Coudé* and *Nasmyth* systems. See also: *star diagonal; Herschel wedge.*

A plane diagonal mirror such as might be used in Newtonian, Nasmyth or Coudé telescopes. (Image reproduced by courtesy of Meade Instrument Corp.)

diamond milling*
See *grinding*.

diamond ring
The last *Baily's bead*. Seen just before a solar eclipse becomes total, the chromosphere is becoming visible and the last portion of the photosphere is still to be seen through a lunar valley. The appearance is thus that of a faint red ring with a dazzlingly bright 'diamond'. (CAUTION – always take appropriate precautions when observing the Sun or damage to the eye and / or telescope can result.)

D

diaphragm
See *stop*.

dichotomy
The moment when a planetary disk is exactly half divided between the illuminated and shadowed portions (equivalent to the first or last quarter Moon). The *terminator* therefore runs along the *central meridian*.

dichroic crystal
A crystal, such as iodized polyvinyl alcohol, that has close to 100% absorption of light polarized in one direction, while transmitting most of the orthogonally polarized light. See also: *Polaroid*.

dichroic mirror
A mirror that reflects at one wavelength and is transparent at another. For example a gold-plated mirror is sometimes used in *infrared* observing since it reflects the infrared, but allows some visual light to pass through. The latter may then be used for *finding* and *guiding*.

Dicke radiometer
A *radio receiver* that switches rapidly between the source and an artificial standard, in order to provide a very stable system. The null-balancing Dicke radiometer adjusts the artificial standard to equal the source being observed.

diffraction
A phenomenon of wave motion whereby a proportion of the wave after passing an obstruction is bent into what should be the shadow zone. Combined with *interference*, diffraction causes the *Airy disk* and its surrounding fringes through the diffraction of light waves at the edges of the telescope *objective*.

diffraction grating
A device that produces a *spectrum* through the effects of *diffraction* and *interference*. Light passes through a number of thin parallel apertures (transmission grating), or is reflected from a number of thin parallel mirrors (reflection grating). The beams from each aperture/mirror are diffracted into each other and constructive interference then produces the spectra. The zero order spectrum has a path difference of zero wavelengths between the beams from successive apertures/mirrors, and is white (i.e. not a spectrum). On either side it has first order spectra where the path difference is one wavelength between the beams from successive apertures/mirrors. The second order spectra resulting from path differences of two wavelengths between the beams from successive apertures/mirrors then appear, thereafter the third, fourth, fifth, etc. orders. The grating must be *blazed* if it is to work efficiently. The *spectral resolution* of a grating is given by Nm, where N is the number of apertures/mirrors and m is the spectral order being used. The spectral *dispersion* increases as the spectral order

increases. For optical work in astronomy reflection gratings are commonly used with sizes up to 100 mm square and with up to 2,000 lines (mirrors) per mm. Diffraction gratings may be plane or curved. In the latter case, relatively high efficiencies can be obtained since the grating can act as its own collimator and/or imaging optics. See also: *Rowland circle.*

diffraction limited optics

Lenses and mirrors of sufficient quality that the resulting instrument is able to reach its *Rayleigh resolution*. For a mirror the surface must be correctly shaped to within an eighth of its operating wavelength, and for lenses the surfaces must be correct to within a quarter of the operating wavelength, in order for the final instrument to be diffraction-limited. These requirements correspond to a deviation of the *wave front* from perfection of no more than a quarter of the wavelength. Since the Rayleigh resolution is an arbitrary definition, which can be improved on by many observers, there may be advantages in having optics corrected to better than the *$\lambda/8$ or $\lambda/4$ criteria*. Radio telescopes often have surface accuracies of $\lambda/20$, for example.

diffraction-limited resolution

See *Rayleigh resolution* and *interferometer resolution.*

Digges' telescope

A possible design of telescope produced by Leonard Digges around 1550, and thus pre-dating its conventionally acknowledged invention by Hans Lippershey in 1608. Digges' telescope, if that is what it was, seems to have had a lens as the objective, and a concavc mirror as thc cycpiccc. See also: *telescope invention.*

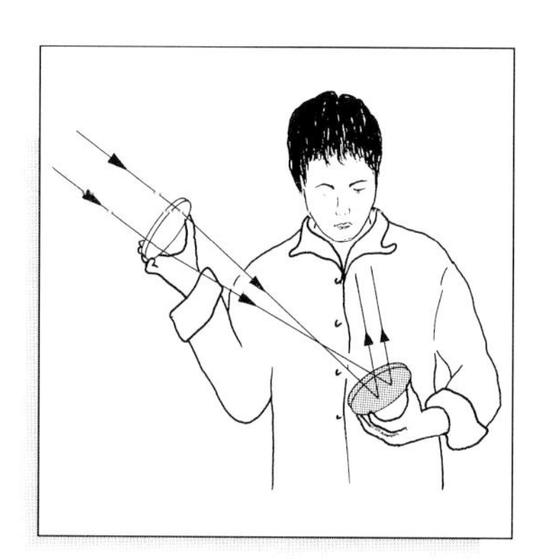

The suggested optical system for the Digges telescope.

digital camera

A still *camera* that uses a *CCD detector.*

digitized sky survey

See *Centre de Données astronomiques de Strasbourg* (SIMBAD), *Lyon-Meudon extragalactic database, NASA extragalactic data set, Palomar observatory sky survey, Sloan digital sky survey* and *STScI digital sky survey.*

dilute aperture

A synonym for un-filled aperture – see *aperture synthesis.*

dioptric system

An optical system based only upon lenses.

dip of the horizon

The excess over 90° of the angle from the *zenith* to a sea-level *horizon* for an observer whose eye is above sea level. It is given by $\theta \approx 0.0296\sqrt{h}$, where θ is the angle of dip in degrees, and h is the height of the observer above sea level in meters.

dipole

See *half-wave dipole.*

direct motion

The 'normal' direction of motion within the solar system. Also known as prograde motion. For planets in the sky, it is from West to East against the background stars, for rotating or orbiting solar system objects it is anti-clockwise as seen from above the solar system in the direction of the *north pole*. See also: *retrograde motion.*

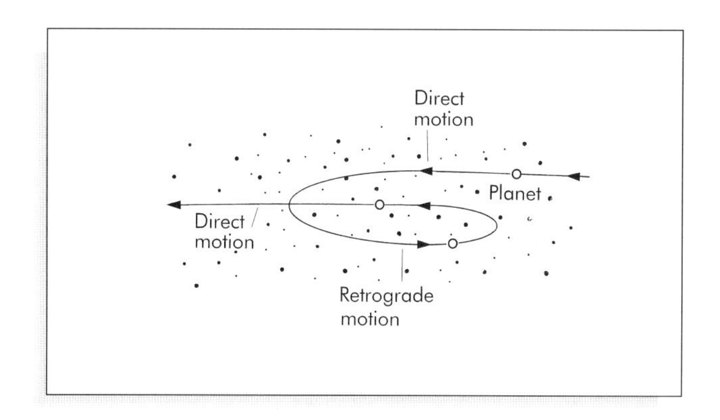

The movement of a planet across the sky (schematic) showing direct and retrograde motion.

direct vision spectroscope

A small *spectroscope* in which the spectrum emerges undeviated (i.e. continuing in the same direction as the entrance beam). They are often used in the laboratory to inspect artificial sources, but can also be placed after an eyepiece on a telescope to allow the *spectra* of the brighter stars and nebulas to be seen directly. They use several *prisms* (often five) formed of two different glass types, and arranged so that the deviations caused by the prisms cancel out, while the *dispersions* do not. It is in effect the inverse of an *achromatic lens.*

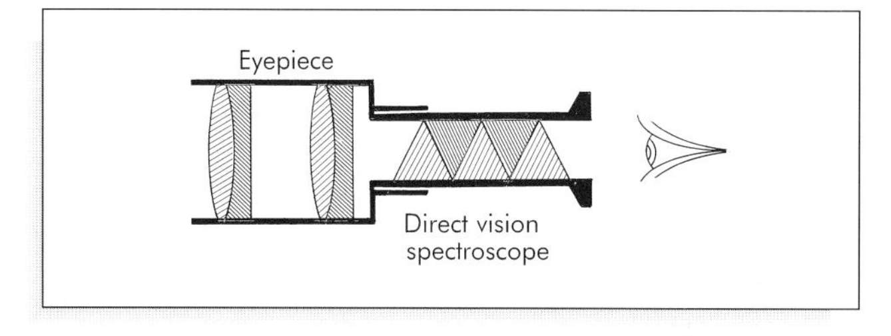

A direct vision spectroscope.

directivity

The *gain* of a perfect *antenna.* Losses in real antennas mean the actual gain is normally less than the directivity.

dirty image

See *raw image.*

Discovery

A commercial firm manufacturing telescopes and accessories.

dish type radio telescope

See *radio telescope.*

dispersion

1 *Refractive index* – see *constringence.*
2 Spectral – The rate of change of wavelength along a *spectrum.* Linear dispersion is the change along the image of the spectrum, and is usually expressed in nm/mm or Å/mm. Angular dispersion is the change with the angle at which the light emerges from the *prism* or *diffraction grating.*

dispersive power

A measure of the *dispersion* of an optical material, equal to the reciprocal of the *constringence.*

distance determination

See *moving cluster method, standard candle method* and *parallax.*

distance modulus

The difference between the *apparent magnitude* and *absolute magnitude* of an object. It is directly related to distance, D (in pc) by:

$$m - M = 5\log_{10}(D) - 5 \qquad \text{or} \qquad D = 10^{0.2(m-M+5)} \quad pc$$

and forms the basis for the *standard candle method* of distance determination.

D

distortion

A fault in images arising from the magnification increasing or decreasing away from the optical axis. The former results in barrel distortion, the latter in pin-cushion distortion. See also: *aberration (optical).*

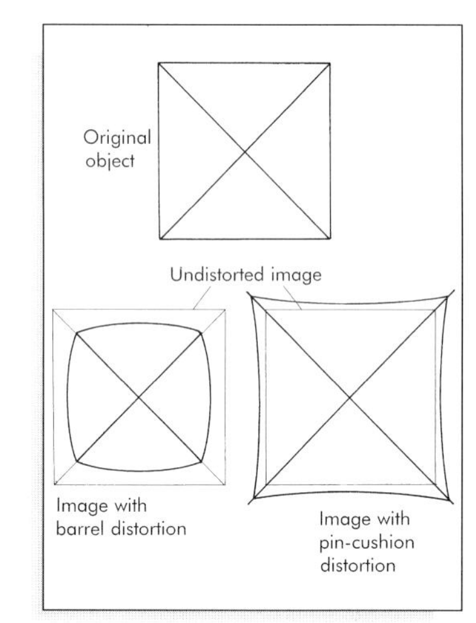

The effect of distortion on the image of an extended source.

dithering

A technique for obtaining images from *CCDs* with a better resolution that that given by the *pixel* size. Several images are obtained with a shift between each of less than the size of a pixel. The images are then combined to give an image with sub-pixel resolution.

diurnal aberration

Aberration arising from the observer's velocity due to the Earth's rotation. At maximum (on the equator) it amounts to only 0.3″.

diurnal inequality

The change in the motion of a nearby object such as the Moon or nearer planets arising from *diurnal parallax.*

diurnal libration

A change in the sections of the lunar surface visible to an observer arising from *diurnal parallax.* Diurnal libration allows the observer to see about 1° more around the eastern or western lunar limbs at Moon rise or Moon set, compared with when the Moon *culminates.* See also: *libration.*

diurnal motion

The daily movement of objects across the sky from east to west arising from the Earth's rotation in the other direction.

diurnal parallax

A change in the position of an object such as the Moon or a nearby planet. Also known as geocentric parallax. The effect is due to the observer's changing position as the Earth rotates, and its precise definition is the difference between the position of the object as seen by the observer and the position as it would be seen from the center of the Earth. Its maximum magnitude, known as the horizontal parallax, since the object is then on the observer's horizon (and which is equal to the angular radius of the Earth as seen from the object) is:

Moon 1° (mean value 57′ 03″, known as lunar parallax)
Sun 8.8″ (known as solar parallax)

Venus 32″
Mars 23″
Jupiter 2.1″.

diverging lens

A *lens* that causes an incident parallel beam of light to diverge from a *virtual image* at its *focus*. Also known as a negative lens. See also: *converging lens.*

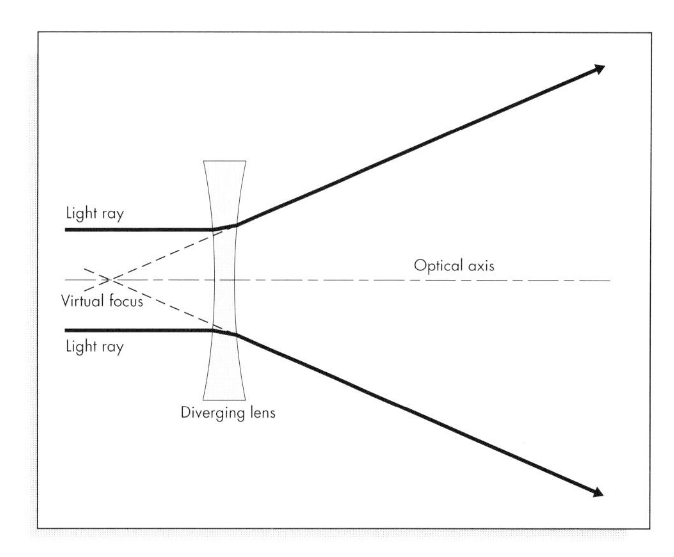

Light paths through a diverging lens.

D

D lines

See *Fraunhofer lines.*

Dobsonian mounting

See panel (page 68).

dodging

A technique used in photographic printing to enable both bright and faint parts of an image to be shown. It involves shading the light part of the negative (the faint part of the original object) during its enlargement. A suitably shaped piece of cardboard is hand held above the printing paper and moved continuously so that there are no sharp shadow edges on the final print. The mask is removed when sufficient of the enlarging exposure is left for that part of the image to register on the print.

Dollond eyepiece

A class C (see *eyepieces*) eyepiece that is a variation of the *Huyghenian eyepiece* in which the *field lens focal length* is three times that of the *eye lens.*

dome

The top, moving part of an *observatory* building. Domes are often hemispherical in shape, and have an aperture (slot) through which the telescope points. The aperture is opened and closed by means of a shutter. The dome rotates in *azimuth* to allow access to all parts of the sky. The 'dome' may sometimes be cylindrical or conical or some other shape if economics or the observing requirements so dictate. This is especially the case with observatories for small telescopes, where the DIY construction problems of a hemispherical structure may outweigh the disadvantages of an alternative shape, and for very large telescopes, especially those on *alt-azimuth* mountings, where another shape may be cheaper to construct. See also: *radome; Lanphier shutter; clam shell dome.*

Denis Buczynski's 23-foot diameter dome.

Dobsonian mounting

A telescope mounting of the *alt–azimuth* type. It is particularly suited to DIY production and can take quite large telescopes. The telescope is supported by two circles, which rest in two semi-circular cut outs in a U-shaped support to allow the telescope to move in *altitude*. The U-shaped cradle then pivots at its center to give motion in *azimuth*. Teflon pads are usually used to give low-friction bearings. See also: *Equatorial platform; Poncet platform.*

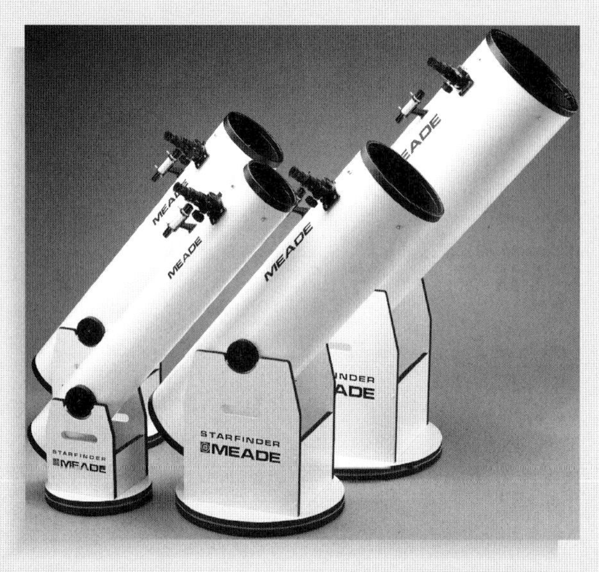

Newtonian telescopes on Dobsonian mountings (Image reproduced by courtesy of Meade Instrument Corp).

David Fredericksen and his 0.31 m (12.5-inch) Newtonian telescope on a Dobsonian mounting.

Tom and Jeannie Clark's 0.9 m (36-inch) Newtonian telescope on a Dobsonian mounting.

Gil Stacy's 0.39 m (15.5-inch) Newtonian telescope on a Dobsonian mounting. The telescope can easily be dismantled for transportation.

Dominion astrophysical observatory
An optical observatory in Victoria, British Columbia, altitude 200 m.

Dominion radio astrophysical observatory
A radio observatory at Penticton, British Columbia. The largest instrument is a 26m dish.

D

Doppler broadening
An increase in the widths of *spectrum lines* arising from Doppler shifts for the atoms or ions emitting or absorbing the line. The velocities may arise from thermal motions, turbulence, rotation of the star, etc.

Doppler shift
The change in the *wavelength* of a *spectrum line* arising from the relative line-of-sight motion between the observer and the source. Wavelengths are increased if the relative motion is away from each other (red-shift, but the principle applies at all wavelengths) and decreased if the relative motion is towards each other (blue-shift). The velocities are then positive and negative respectively. The velocity, v, is related to the Doppler shift, $\Delta\lambda$, by:

$$\frac{v}{c} = \frac{\Delta\lambda}{\lambda} = \frac{\lambda_{observed} - \lambda_{laboratory}}{\lambda_{laboratory}},$$

where c is the velocity of light in a vacuum (3×10^8 m/s).

double star
A pair of stars with a small angular *separation*. The stars may simply happen to lie nearly along the same line of sight but actually be at very different distances from the Earth, or they may be close together in space and gravitationally linked. In the latter case they are termed a *binary star*. The brighter star is termed the primary, and the fainter, the secondary or comes (Latin for "companion"). The position angle is the angle from the North of the secondary from the primary in the sense N–E–S–W.

doublet
1 A lens formed from two components, such as the *achromat*.
2 A pair of close *spectrum lines* arising from the same atom, ion or molecule and with a common *energy level*.

DQE
See *detective quantum efficiency*.

drain
See *anti-blooming*.

draw tube
A pair of nested tubes on the end of a telescope into which is mounted the *eyepiece*, and which allows focusing by sliding one tube up and down inside the other. See also: *eyepiece mount; focuser*.

drift scan
A technique whereby an image is built up using a non-imaging *detector*. The detector drifts across the image due to the motion of the source and/or of the detector/telescope. The varying output from the detector then corresponds to a transect across the image. Moving the detector orthogonally to the drift motion gives the next transect,

and allows a two-dimensional image to be reconstructed by a raster scan like that on a television. The drift motion may be due to the diurnal rotation of the Earth, or to the rotation in space of a rocket or spacecraft, etc.

D

drive

The mechanical arrangement for moving a telescope to allow it to *track* objects and for *finding* and *guiding* on objects. On an *equatorial mount*, only a single constant speed motor is required to rotate the telescope around the *polar axis* in order to track objects. Other types of mounting require motors for both axes whose speeds usually need to be variable for tracking. All types of mounting need motors on all axes for finding and guiding. It is convenient if several pre-set speeds are available on the drive motors to allow rapid movement across large angles for finding, and a much slower movement, giving fine control, for guiding.

DUMAND neutrino detector

A *neutrino detector*, based upon detecting the Čerenkov radiation from high-energy electrons produced in water within the Pacific ocean.

dusk

See *twilight*.

Dwingeloo radio observatory

A radio observatory at Dwingeloo, Netherlands, with a 25m dish telescope.

dynamical equinox

The zero point for *right ascension* at a given moment of time. It moves due to *precession*, and catalogued star positions must be corrected for the movement since the *epoch* of the catalog in order to give actual positions.

dynamic parallax

The *parallax* of a visual *binary star* determined by measuring the angular semi-major axis of the orbit directly and inferring its physical size using Kepler's third law.

dynamic range

The range of a *detector* between the minimum and maximum detectable signals. The minimum signal is usually determined by the *noise* of the system, and the maximum when the detector *saturates*. For a *photographic emulsion* the dynamic range may be about ×1,000, while for a *CCD* it can be ×500,000. See also: *signal/noise ratio*.

dynode

An intermediate electrode, especially within a *photomultiplier*.

Earth rotation synthesis

A synonym for *aperture synthesis.*

Earthshine

The illumination of the dark side of the Moon by reflected light from the Earth. See also: *Old Moon in the New Moon's arms.*

eastern elongation

The *elongation* when an object is to the east of the Sun (i.e. it sets after the Sun). See also: *western elongation; greatest elongation; aspect.*

eccentricity

A measure of the shape of a *conic section*. It takes the values:

circle $e = 0$
ellipse $0 < e < 1$
parabola $e = 1$
hyperbola $e > 1$.

For ellipses, which represent the shapes of most orbits, the value of e is given by:

$$e = \sqrt{1 - \frac{b^2}{a^2}},$$

where a is the semi-major axis of the ellipse (or orbit) and b the semi-minor axis. The major planets have orbits whose eccentricities are close to zero. Pluto has the highest value at 0.25, followed by Mercury at 0.206 and Mars at 0.093. The eccentricity of the Earth's orbit is 0.017, and Venus has the most circular orbit with a value of e of 0.007. Minor planets have a much wider range of values of e, but all still less than 1. Many comets though have orbits with eccentricities close to 1 (i.e. parabolic orbits). See also: *oblateness.*

echelle grating

A very coarse *diffraction grating* that is illuminated at a small angle to the plane of the grating. The grating produces many high order *spectra* with high *dispersion* and *resolution*. But the numerous high order spectra overlap, and must be separated by a *cross disperser*. See also: *Fabry–Perot etalon.*

eclipse

See panel (page 72).

eclipse spectacles

See *solar eclipse viewer.*

ecliptic

The path of the Sun against the background stars over a period of a year. Also, the extension of the plane of the Earth's orbit to meet the *celestial sphere*. The ecliptic is the center line of the *zodiac*. See also: *obliquity of the ecliptic.*

ecliptic latitude

See *celestial latitude.*

eclipse

The passage of one object in the sky in front of another, when the two objects are of roughly similar angular sizes. For example, a *solar eclipse*, when the Moon passes in front of the Sun, or an eclipsing *binary star*, when one star of the system passes in front of the other. Note that *lunar eclipses* are not strictly eclipses, since the Moon is passing into the shadow of the Earth, although the phenomenon would be genuine solar eclipse for an observer on the Moon. (Solar and lunar eclipses are tabulated annually in the Astronomical *Almanac*.) See also: *occultation; transit.*

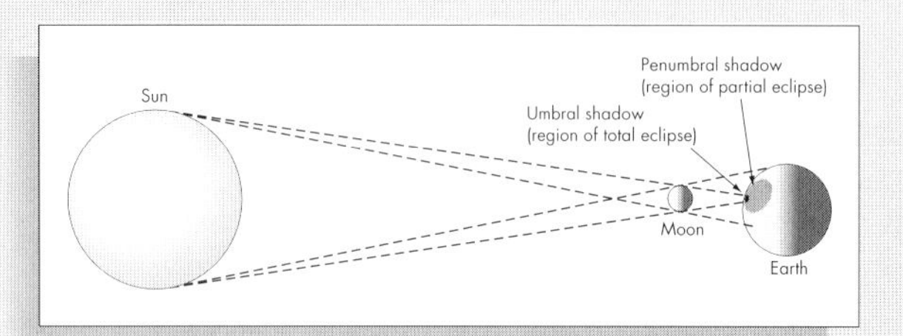

The geometry of a solar eclipse (not to scale).

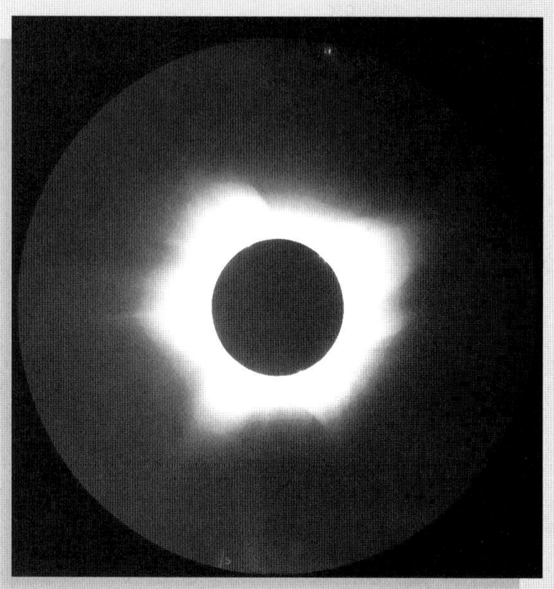

The solar corona during a total solar eclipse (Image reproduced by courtesy of the Royal Astronomical Society).

Eyepiece projection of a partial solar eclipse.

Solar prominences and the inner part of the corona during a total solar eclipse (Image reproduced by courtesy of the Royal Astronomical Society).

ecliptic longitude

See *celestial longitude.*

ecliptic pole

The points on the celestial sphere orthogonal to the plane of the *ecliptic*. They are at RA 18h Dec +66° 33′ (midway between δ and ζ Draconis) and RA 6h Dec −66° 33′ (near δ Doradus). The *north* and *south poles* move in circles centered on the ecliptic poles with a period of 25,800 years due to *precession.*

E

edge effect

An effect in photographic images whereby near the boundary of two areas of high contrast, the dark area will be overdeveloped (i.e. too dark), and the light area will be underdeveloped (too light). The effect arises through the diffusion of unused developer through the emulsion from light area to the dark area.

edge enhancement

An *image-processing filter* that emphasizes parts of the image where the intensity is changing rapidly.

effective aperture

1 For a *radio telescope* the ratio between the power delivered to the *receiver* and the power per unit area from the source at the *antenna*. It is also known as the effective area, and is the aperture of a 100% efficient dish-type radio telescope that would deliver the same power to the receiver. Actual dish type radio telescopes have effective apertures smaller than their real apertures, because not all the energy reflected by the dish is converted into a signal by the antenna. The effective aperture of a *Yagi antenna* is approximately given by the square of the operating wavelength. Effective aperture, A_e, and *gain*, g, are related by:

$$A_e = \frac{g\lambda^2}{4\pi}.$$

2 For any telescope, the aperture of a 100% efficient instrument that would deliver the same total energy to the detector. The effective apertures of real telescopes are always less than their true apertures because of vignetting, absorption and reflection losses in the optics, and obscuration by *secondary mirrors*, mirror supports, etc.

effective area

See *effective aperture.*

effective focal length

The focal length of a simple telescope with the same diameter *objective* and the same *image scale* as the telescope actually being used. The effective focal length differs from the intrinsic *focal length* of the objective because of the effects of *secondary mirrors, Barlow lenses, tele-compressors*, etc. which expand or contract the converging light beam from the objective. Thus a large modern *Ritchey–Chrétien telescope* might have a primary mirror with a *focal ratio* of f2 to f3, but a final focal ratio of f8 to f12, as a result of the amplifying effect of its secondary mirror on its focal length. Similarly, if a Barlow lens is used with an eyepiece on a small telescope to increase the magnification by a factor of 2 or 3, then the effective focal length of the telescope is increased by the same factor.

When an eyepiece is used for *eyepiece projection;* the effective focal length, f_{EFL}, of the telescope is:

$$f_{EFL} \approx f_o \times \frac{d}{f_e},$$

where f_o is the focal length of the objective, f_e is the focal length of the eyepiece, and d is the projection distance from the eyepiece. See also: *equivalent focal length.*

effective focal ratio

The *focal ratio* calculated using the *effective focal length* of the instrument in place of the actual *focal length.*

effective temperature

The temperature of a *black body* that would radiate the same total amount of energy per unit area as the object concerned. Also known as brightness temperature. For most purposes this is the most representative temperature, but other methods of measuring temperature based on thermal motions, levels of excitation and ionization may give very different results, and need to be used in special applications. For example the *kinetic temperature* of the solar corona is about 10^6 K, but its effective temperature is around 100 K, the difference arising because of the low density of the corona. All temperature measures are equal to each other in thermodynamic equilibrium, which is the physical condition to be found deep inside a star. See also: *excitation temperature; ionization temperature.*

Effelsberg radio telescope

A fully steerable dish-type *radio telescope* with a diameter of 100 m, located in the Eifel mountains in Germany. It is the second largest such instrument after the *Green Bank radio telescope.*

e-folding time

The time interval over which the *amplitude* of a cyclic variation changes by a factor of e (= 2.71828 ...).

electric focuser

See *focuser.*

Electrim Co.

A commercial firm manufacturing *CCD* cameras for astronomical use.

electro-magnetic radiation

A phenomenon transporting energy, that is formed from an electric field and a magnetic field which are orthogonal to each other and whose amplitudes vary in a sinusoidal fashion. The energy can also behave as though divided into individual particles (photons or quanta). This wave–particle duality requires quantum mechanics for its complete description. The *wavelength*, λ, of electro-magnetic radiation is the linear distance between successive maxima (or minima) of its electric or magnetic components, and its *frequency*, ν, is the number of maxima passing a given point per second. The two quantities are related by:

$$\lambda \nu = \frac{c}{\mu},$$

where c is the velocity of light in a vacuum (2.9979×10^8 m/s) and m is the *refractive index* of the material through which the radiation is passing (= 1 for a vacuum). The energy of a photon is given by:

$$E = h\nu,$$

where h is Planck's constant and has a value of 6.625×10^{-34} J s (see also *electron volt*, or eV).

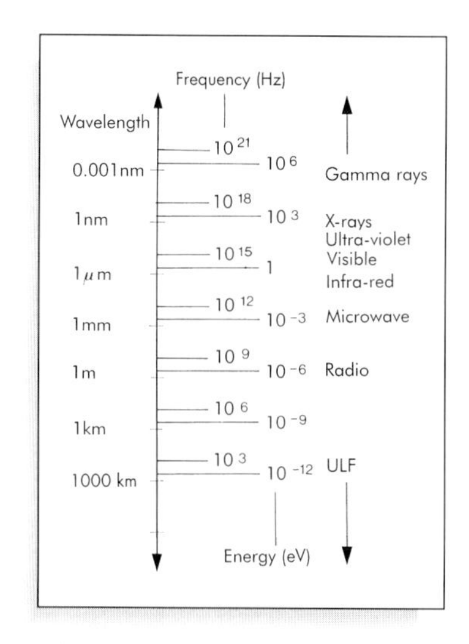

The range of electro-magnetic radiation.

Light and radio waves are the most familiar examples of electro-magnetic radiation, but the complete *spectrum* extends from zero wavelength (infinite frequency and energy) to infinite wavelength (zero frequency and energy). For convenience the different regions of the spectrum are given different names. This is, however, an arbitrary division since the phenomenon is the same at all wavelengths and energies, although the way the electro-magnetic radiation interacts with matter does vary with wavelength. The divisions of the spectrum are:

Region	Wavelength	Frequency (Hz)
gamma rays	0–0.01 nm	∞–3×10^{19}
X-rays	0.01–1 nm	3×10^{19}–3×10^{17}
ultraviolet	1–350 nm	3×10^{17}–8.6×10^{14}
visible	350–700 nm	8.6×10^{14}–4.3×10^{14}
infrared	700 nm–100 μm	4.3×10^{14}–3×10^{12}
microwave	100 μm–10 mm	3×10^{12}–3×10^{10}
radio	10 mm–∞	3×10^{10} – 0

electron

A sub-atomic particle that is a fundamental component of an atom. It has a mass of 9.109×10^{-31} kg, and a negative charge of 1.602×10^{-19} C. The separation of an electron from its atom (production of an electron–hole pair) is the basis of the operation of many *detectors*. See also: *proton; neutron.*

electronic image format

A coding for the layout of data for an image stored in electronic form. There are many varieties of format. One in widespread astronomical use is FITS (Flexible Image Transport System). This has a header that is 2880 bytes in size, giving the size of the image and other details such as whether it has one, two or three dimensions, details of the observation, etc. The middle section is simply the image data in the form specified by the header, while the end section is a number of zeros to pad the whole out to be a multiple of 2880 bytes. Other commonly encountered formats are GIF (compressed, 256 colors only, often used on the internet), JPEG (highly compressed with some loss of information, but small files) and TIFF (can be compressed or uncompressed, true colors, widely used in publishing).

electronic transition

See *transition.*

electronographic camera

A camera based upon an *image intensifier*, in which the electrons pass through a very thin window of mica at the end of the device to produce the image directly in photographic emulsion pressed against the window. The device gave a speed gain of about x10 over normal photographic emulsion, but has now been superseded by the *CCD.*

electron volt

A non-SI unit of energy that is in frequent use, symbol, eV. It is the energy gained by an electron in passing through a potential difference (voltage) of 1 volt, and equals 1.602×10^{-19} J. It is a convenient unit for dealing with sub-atomic particle interactions and the energies of *photons*. Visual photons for example have energies from 1.8 eV (red) to 3.5 eV (violet).

elements

See *orbital elements.*

ellipse

One of the *conic sections*, which represents the shape of most orbits, and which forms the shape of some telescope mirrors. The longest diameter is called the major axis (and the longest radius, the semi-major axis; symbol, a), while the shortest diameter is the minor axis (and the shortest radius, the semi-minor axis; symbol, b). An ellipse has two *foci* that are positioned on the major axis at a distance ae from the center of the ellipse (where e is the *eccentricity* of the ellipse). In a planetary orbit the Sun occupies one of the foci of the elliptical orbit of the planet (Kepler's first law). An elliptical mirror will focus an object placed at one focus on to the other focus. See also: *Gregorian telescope.*

E

elliptical aberration
A small component (0.3″) of stellar *aberration* arising from the elliptical shape of the Earth's orbit.

elliptically polarized radiation
Electromagnetic radiation in which the direction of the electric vector rotates and its amplitude varies with time, at the frequency of the radiation. See also: *circularly polarized radiation; linearly polarized radiation.*

ellipticity
See *oblateness.*

elongation
The angular separation in the sky between an object, especially a planet, and the Sun. It is the Sun–Earth–Object angle. See also: *aspect; greatest elongation; eastern elongation; western elongation.*

emersion
The re-appearance of an object after *occultation* or *eclipse.*

emission line
A bright line appearing in the *spectrum* of an object. The line arises from the emission of *photons* at a specific *wavelength* by ions, atoms or molecules. See also: *absorption line.*

emissivity
The efficiency with which an object radiates energy compared with a *black body.*

e-m radiation
See *electro-magnetic radiation.*

emulsion
See *photographic emulsion.*

energy level
The energy of an electron in an atom, ion or molecule. Also the energy of a vibrating or rotating molecule. Energy levels can take only certain allowed values (known as quantization). When an electron moves from one energy level to another it emits or absorbs a *photon.* Since the energy levels are fixed, so are the energy differences between them and so the photons are emitted or absorbed at a fixed energy or wavelength, giving rise to the *spectrum lines.* The lowest energy level is called the ground state. See also: *Grotrian diagram.*

energy level diagram
See *Grotrian diagram.*

English mounting

A design of *equatorial mounting* for a telescope in which the telescope swings between a rectangular frame (giving the *declination axis*), and that frame is supported on pivots at each end to provide the *polar axis*. Also known as the yoke mounting.

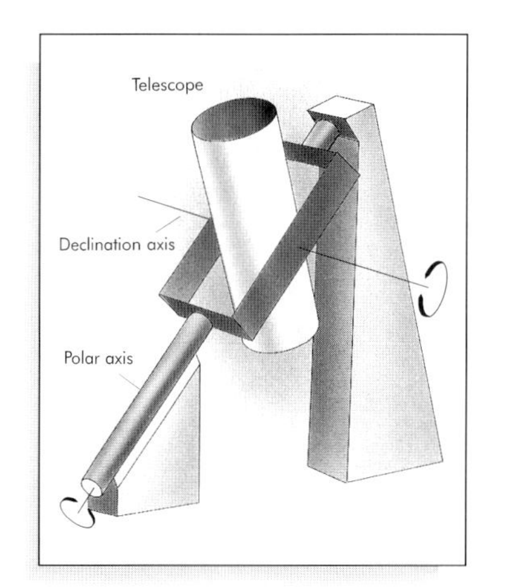

The English, or Yoke, telescope mounting.

enlarging

Any process whereby the size of an image is increased. It is especially applied to *printing* a photographic *negative* on a larger scale than the original. In *image processing* packages the process is often called zooming.

ephemeris

A listing of the predicted positions of a moving object in the sky such as the Moon, Sun, major and minor planets, and comets. See also: *almanac*.

ephemeris time

The basis of time keeping until 1967 when it was replaced by *international atomic time*. Ephemeris time continued in use until 1984 for the purposes of calculating orbital motions within the solar system, when it was replaced by terrestrial time, which is international atomic time plus 32.18 seconds. Ephemeris time was based upon the Earth's orbital motion and the second was defined as 1/31,556,925.9747 of the tropical year for 1900 January 0^{d} 12^{h}. See also: *coordinated universal time*.

epoch

1 The date for which the positions in a *star catalog*, *star atlas*, *ephemeris*, etc. are correct. It is usually midday on 1 January. For ephemerides it is likely to be the current year, while for star catalogs and atlases, it is usually one of a standardized dates at 25-year intervals, e.g. 1 Jan 1950, 1 Jan 1975, or 1 Jan 2000, etc. For dates other than the epoch, the positions will need to be corrected for *precession*.
2 One of the *orbital elements*.

Epsilon telescopes

A range of telescopes produced by the *Takahashi Co*.

equation of light

See *heliocentric time*.

equation of the equinoxes

The difference between *apparent* and *mean sidereal times*, also the *nutation* in *right ascension*. See also: *sidereal time*.

equation of time

The difference between *solar time* and *local mean solar time*. The position of the Sun in the sky at civil midday (or any other fixed civil time, ignoring summer time), varies either side of the meridian. This variation, when combined with the Sun's motion in declination, results in a figure-of-eight shape known as the Analemma, often to be found on antique maps and globes.

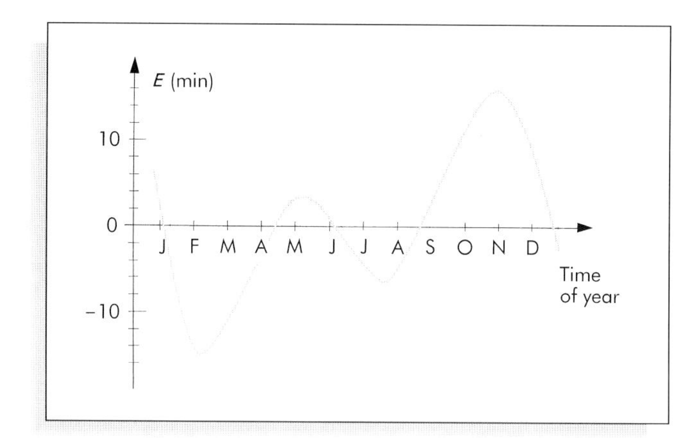

The equation of time.

equatorial platform

A device that allows an *alt-azimuth-mounted* telescope (especially a *Dobsonian*) to track objects using a single, constant velocity motor. The whole telescope is mounted onto a horizontal platform, that in turn moves on an inclined plane set at an angle of (90° – latitude) to the horizontal about a fixed pivot. See also: *Poncet platform.*

Telescope and mounting placed on top plate
L metres
Rate of motion – 0.07292 L sin φ mm per second
Universal joint pivot
Motor
Screw drive
Swivelling wheels (on underside of top plate)
Inclined plane – angle of inclination 90° – φ where φ is the observer's latitude
North

Semi-exploded view of an inclined-plane equatorial drive platform

equatorial telescope mounting

See panel (pages 80–81).

equinox

One of two points in the sky where the *ecliptic* and the *celestial equator* cross each other. See also: *First Point of Aries; autumnal equinox.*

equivalent focal length

For a compound optical system (such as an eyepiece) the focal length of a simple lens that would provide the same result (magnification, focus position, etc.). See also: *effective focal length.*

equivalent width

A measure of the strength of a *spectrum line*. It is the width of a rectangular section of the nearby *continuum* that has the same area as that enclosed by the line.

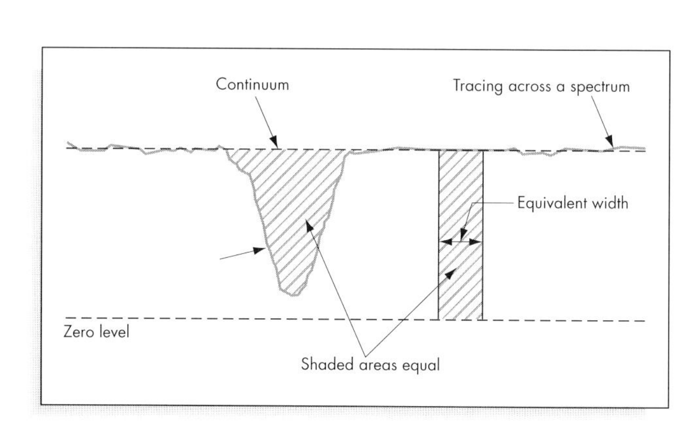

The equivalent width of a spectrum line.

erect image

An upright image, i.e. one with the same orientation as the object seen directly. *Galilean* and *Gregorian telescopes* produce erect images, but more normal astronomical telescope designs produce an *inverted image*, and so need a *relay lens* if they are to be used for terrestrial observing.

erecting prism

A prism incorporated into the light path of a telescope or pair of binoculars in order to give an *erect image*. For example, two opposed 90° prisms in which the light paths enter and exit through their long faces are used in many *binoculars*. The combination is known as a *Porro prism* and also acts to shorten the length of the instrument by folding the light beams. The *penta-prism* in a single-lens reflex camera acts in a similar manner.

E

erector

See *erecting prism*.

Erfle eyepiece

A class A (see *eyepieces*) eyepiece that is made from two *achromatic lenses* plus a third lens that may be simple or achromatic. The design has the wide useable *field of view* of 60° to 70°, and good *eye relief*. It suffers, however, from astigmatism, and so is best suited to telescopes with focal ratios of f6 or slower.

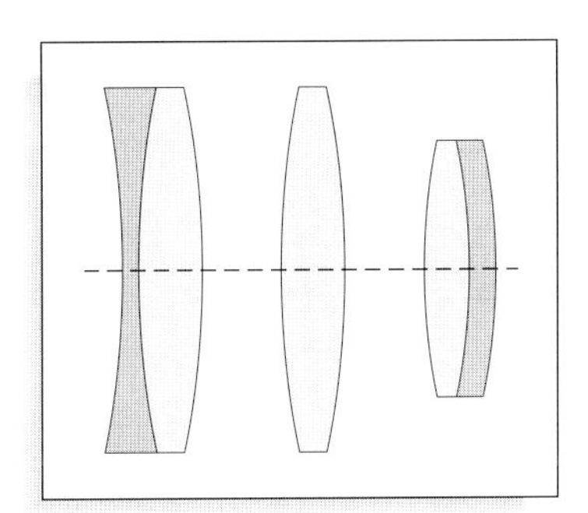

The optical arrangement of the Erfle eyepiece.

error box

An area of the sky, usually but not necessarily rectangular in shape, within which a source is likely to be found. Error boxes are used mostly for X-ray and gamma ray observations, and the source will have a certain probability (often 95% or 99%) of being found within the box based upon the known angular resolution and pointing accuracy of the instrument.

errors

See *uncertainties*.

ESA

See *European space agency*.

ESO

See *European Southern Observatory*.

etalon

See *Fabry–Perot etalon*.

etendu

A synonym for *throughput*.

equatorial telescope mounting

One of several related designs of mountings for telescopes that are in widespread use today, and which were used almost universally until a couple of decades ago. All the variants have one axis of rotation, the *polar axis*, which is parallel with the Earth's rotational axis. The telescope can thus *track* objects by being driven around the polar axis in the opposite direction to the Earth's rotation at a constant velocity of one revolution per *sidereal day*. *Hour angle* can be read off directly from a *setting circle* attached to the polar axis. *Right ascension* may be read from a setting circle that also rotates once per sidereal day but independently of the telescope. Access to the whole sky is provided by motion around a second axis that is orthogonal to the polar axis. This is called the *declination axis*, and a setting circle attached to it reads *declination* directly. See also: *alt–az mounting; English mounting; modified English mounting; fork mounting; German mounting; horseshoe mounting; wedge.*

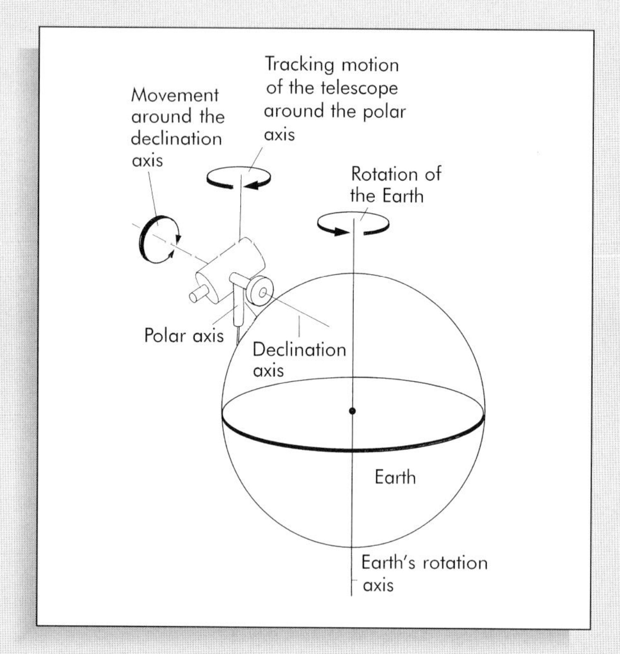

The alignment of the axes of an equatorial telescope mounting with the Earth.

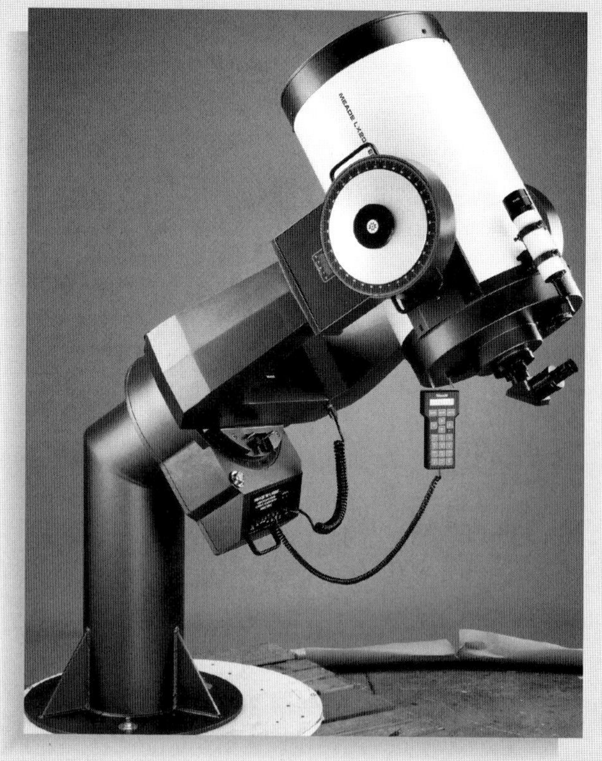

A massive equatorial mounting for a Meade 0.4 m (16-inch) LX 200 telescope (Image reproduced by courtesy of Meade Instrument Corp).

A simple means of converting an al-azimuth mount for an ETX telescope to an equatorial mounting (Image reproduced by courtesy of Meade Instrument Corp).

Martin Mobberley and the equatorial mounting for a small telescope that he made at the age of 15.

E

ETX telescope

A range of small telescopes of the *Maksutov* design produced by the *Meade Instrument corporation.*

The Meade ETX telescope. (Image reproduced by courtesy of Meade Instrument Corp.)

European Southern Observatory

An observatory operated jointly by a number of European countries at *La Silla* and *Cerro Paranal* in the Atacama desert, Chile, altitude 2500 m. The headquarters are at Garching near Munich, Germany. Its instrumentation includes the *very large telescope*, the 3.5m new technology telescope, a 3.6m telescope and a 1m *Schmidt camera*, plus several other telescopes. See homepage at http://www.eso.org/.

The European Southern Observatory site at Cerro Pachon. (Image reproduced by courtesy of the European Southern Observatory.)

European Space Agency

A consortium of a number of European countries, including the UK, whose objective is to develop rockets and spacecraft, their associated technology, their ground-based facilities and to conduct space research. Its headquarters are in Paris, and the launch site is at Kourou, French Guiana. The main launch vehicles are the Ariane series of rockets. See homepage at http://esa.int/.

EUV

See *extreme ultraviolet region.*

eV

See *electron volt.*

evening star

Venus around the time of *greatest brilliance*, and when its *elongation* is east of the Sun, so that it is visible in the evening sky after sunset.

excitation

The process of an electron in an atom, ion or molecule gaining energy from a *photon*, collision or other source, and moving to a higher *energy level*. When a photon supplies the energy for excitation, the resulting deficit of photons at that energy results in *absorption lines* in the *spectrum*.

excitation potential

The energy required to raise an electron from the *ground state* of an atom or ion to a higher *energy level.*

excitation temperature

A temperature based upon the numbers of electrons in different energy levels within atoms or ions. The excitation temperature appears within Boltzmann's equation:

$$\frac{N_b}{N_a} = \frac{g_b}{g_a} e^{-(E_b - E_a)/kT},$$

where N_a and N_b are the numbers (or number densities) of electrons in energy levels a and b of the atom or ion, E_a and E_b are their respective *excitation potentials*, g_a and g_b are the statistical weights of the levels (the probabilities that the levels will be occupied by an electron, usually a number between 1 and 10) and k is Boltzmann's constant ($= 1.38062 \times 10^{-23}$ J K^{-1}).

E

exit pupil

The smallest area behind an *eyepiece* through which all the light gathered by the telescope passes, also known as the Ramsden disk. It is the image of the *objective* produced by the eyepiece, and it should coincide with the pupil of the eye for optimum viewing. Its diameter is the diameter of the *objective* divided by the *magnification*. See also: *minimum magnification*.

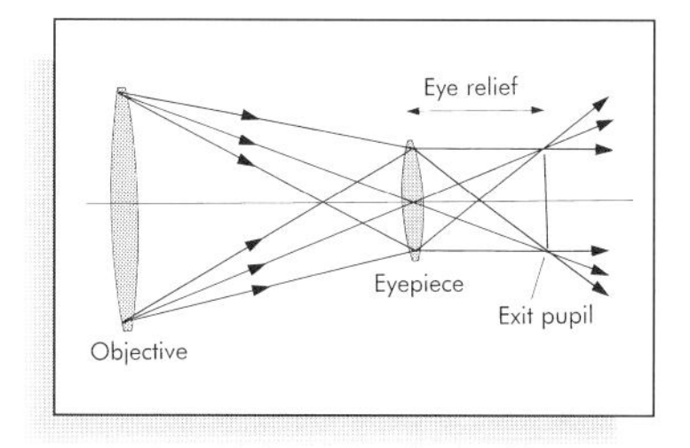

The exit pupil of an eyepiece.

exposure

The length of time that an image is allowed to fall onto a detector, especially a *CCD* or *photographic emulsion*. For astronomical imaging, exposure lengths range from fractions of a second to many hours. See also: *top hat exposure*.

extended source

An object whose image is larger than the *resolution* of the telescope and/or which extends over more than one *pixel* on the detector. The telescope does not increase the surface brightness of an extended object since the increased light gathered by the telescope is spread out over a larger area. For *magnifications* greater than the *minimum*, the surface brightness will actually be less than that to the unaided eye. Hybrid objects that are initially *point sources*, but after a certain level of magnification become extended sources will increase in brightness but by less than the *light grasp*.

extinction

See *atmospheric extinction* and *interstellar absorption*.

extraordinary ray

See *birefringence*.

extrapolation

The extension of the trend a graph (or other data set) beyond its final data point to predict a new value. See also: *interpolation*.

extreme ultraviolet region

The short wavelength end of the *ultraviolet* region, abbreviation, EUV, and covering approximately the *wavelength* region, 1 nm to 100 nm. Also called the XUV region (X-ray/ultraviolet). See also: *electro-magnetic radiation*.

eye
The organ of vision. The main optical components are a simple lens at the front of the eye, preceded by a variable diaphragm (the *eye's pupil*), and the cornea, and the detecting layer at the rear of the eye (see *retina*).

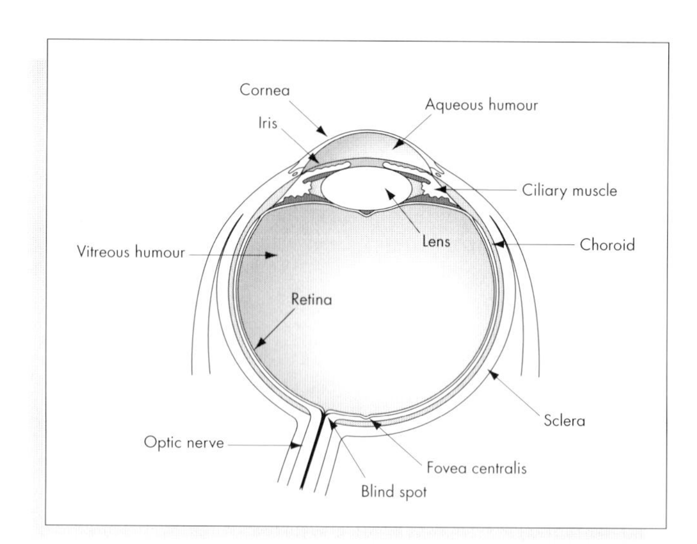

A horizontal section through the human eye, showing its structure.

eye lens
The lens of an *eyepiece* nearest to the observer's eye. See also: *field lens.*

eyepiece
See panel (pages 85–86).

eyepiece mount
A device to hold the *eyepiece* in the correct position at the end of the telescope and to allow the eyepiece to move back and forth along the *optical axis* in order to *focus* the image. It may be a simple *drawtube*, or a more elaborate construction with a geared or screw focusing drive, and a read-out of the eyepiece position. It may also be provided with a rotating carousel into which several eyepieces may be mounted, and so inter-changed rapidly. In this latter case, it is useful if the eyepieces are *par-focal.*

Most modern eyepieces are a push fit into the eyepiece mounting, with standard tube diameters of 1.25″ (31.75 mm) or 2″ (50.8 mm). Eyepieces produced for microscopes, however, are likely to have a barrel diameter of 0.9″ (22.9 mm). While older telescope may still use the RAS thread. This latter is a pre-metric thread designed for gas pipe fittings and adapted for astronomical use. It has a pitch of 0.0625" and a diameter of 1.25″. See also: *focuser.*

eyepiece projection
A setting of the eyepiece that produces a real *image* some distance behind the eyepiece. It is often used to enable safe observation of the Sun, and to provide increased *magnification* of the image on a *detector*.

CAUTION – always take appropriate precautions when observing the Sun or damage to the eye and/or telescope can result. Some telescopes are unsuited to projection of the solar image, and the manufacturer's guarantee may be invalidated. Always check with the manufacturer of your telescope. See also: *solar observing; solar telescope.*

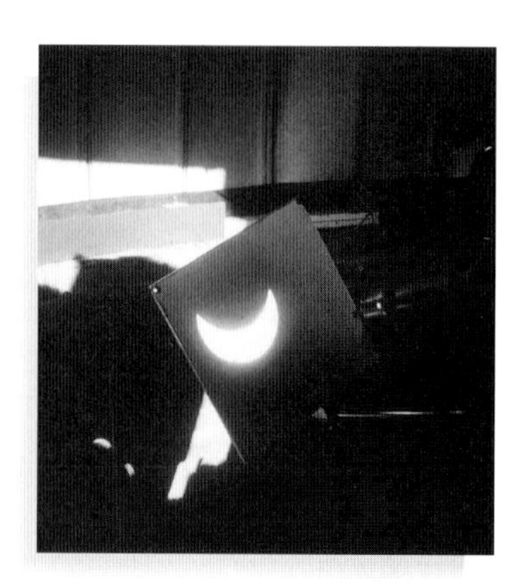

Eyepiece projection of the partially eclipsed Sun.

eyepiece

An optical device based on one or more lenses that produces the visible image from a telescope, also known as an ocular. The *magnification* of the image depends on the *focal lengths* of the eyepiece and the *objective;* There are numerous designs of eyepiece, the differences between them are largely their level of correction of *aberrations*, their *fields of view* and their *eye-relief*. They may be divided into three groups on the basis of their cost and performance.

At the bottom end there are class C designs such as the *Dollond*, *Ramsden* and *Huyghenian*. These have small fields of view and may suffer from other aberrations such as *field curvature*. Though cheap, they are generally not worth considering.

In the middle range, class B, there are eyepieces such as the *Kellner*, *orthoscopic*, *monocentric*, *triple loupe* and *Plössl*. These designs are well corrected for aberrations, have a wider field of view (45° to 65°) and are only a little more expensive than the class C eyepieces. They are the eyepieces of choice for most people for most purposes.

The most expensive group, class A, are the wide-angle eyepieces, such as the *König*, *panoptic*, *ultra-wide angle*, *lanthanum; Erfle* and *Nagler*. Their fields of view can be up to 85° (note that the field of view of the eye is about 50°). There is little point in purchasing a short focal length wide angle eyepiece, since high magnifications will be used for examining fine details with the image normally centered in the field of view, and a class B eyepiece will perform equally well for such on-axis images. It is worth considering a low *power* wide-angle eyepiece, however, for *finding*, and for viewing large *extended objects*. A zoom eyepiece, whose focal length is variable, may also be used, but its optical performance will normally be poorer than a fixed focus eyepiece, though it provides a very convenient means of a quick look at objects. See also: *Barlow lens; sweet spot.*

Eyepieces (Image reproduced by courtesy of Meade Instrument Corp).

An extra-wide angle eyepiece compared with a more conventional one.

An illuminated cross-wire eyepiece for guiding (Image reproduced by courtesy of Meade Instrument Corp).

Eyepieces are expensive – it's worth making a special container to store them safely.

eye relief

The distance between the *eye lens* of an eyepiece and its *exit pupil*. For comfortable viewing it should be between about 6 mm and 10 mm or up to 20 mm if you wear spectacles.

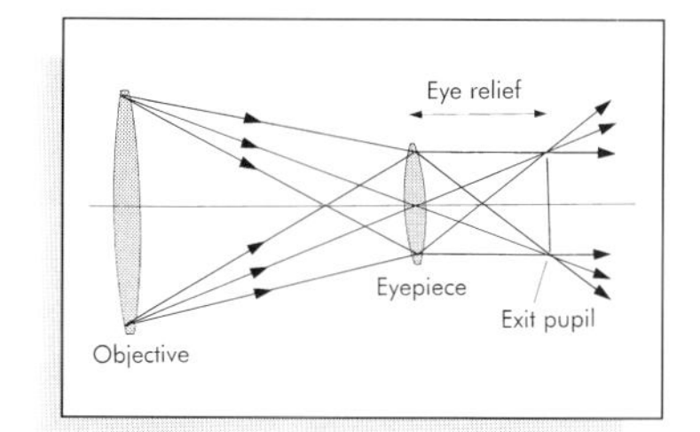

The eye relief of an eyepiece.

eye's pupil

The entrance aperture to the *eye*. It is variable in diameter from about 2 mm under bright illumination to about 7 mm in the dark. See also: *dark adaptation; exit pupil.*

E

F

Fabry lens

A lens placed in front of a detector that images the *objective* onto the detector. The image of the objective is a uniformly illuminated circle that does not move as the image of the object moves due to *seeing* or to *tracking* errors. The use of a Fabry lens therefore reduces the *noise* arising from such sources. Its use is also advantageous when using a detector, such as a *photomultiplier*, whose sensitivity may vary over the detector area.

Fabry–Perot etalon

A device for producing high *dispersion*, high-resolution *spectra*. It comprises two partially reflecting plane mirrors held a short distance apart and aligned accurately parallel with each other. *Interference* among the multiple reflections between the mirrors leads to the production of many overlapping high order segments of a spectrum. A *cross disperser* is then used to separate the segments.

It is also the basis of interference filters. In these, the separation of the mirrors is small (100 nm to 500 nm) and the gap is filled with a transparent material such as magnesium fluoride. In combination with a *band-pass filter* to eliminate unwanted transmission bands, such a device will then have a *bandwidth* ranging from fractions of a nanometer to a few nanometers. The central wavelength and the bandwidth can both be tuned by altering the separation. *H-α filters* are an example of this application, where the central wavelength is 656.28 nm, and the bandwidth in the region of 0.05 nm. See also: *echelle grating*.

false color image

An *image processing* technique whereby a color image is generated with the wavelengths of the original monochromatic images different from the wavelengths at which they are reproduced. For example, monochromatic images obtained in the red, near infrared and medium infrared, might be reproduced as blue, green and red respectively in the false color image. See also: *color balance*.

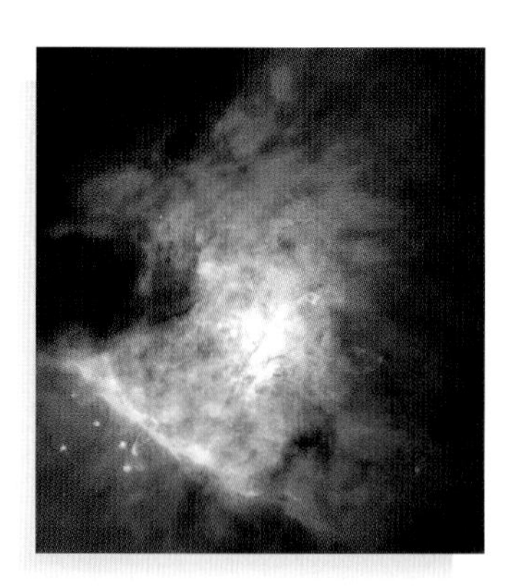

a An image of the Orion nebula taken using the European Southern Observatory's VLT in the near infrared and displayed as a false-color image. (Image reproduced by courtesy of the European Southern Observatory).

b A normal optical image for comparison. (Image reproduced by courtesy of NASA/JPL/Caltech.)

false dawn

The *zodiacal light*, which appears as a brightening of the sky on the horizon where the Sun will rise in due course.

false observation

An observation in which the observer sees or thinks he/she sees something that is not there. This may be due to pre-conceived ideas, for example the rotation of Mercury was thought to be tidally locked onto the Sun and for many years observations seemed to bear this out, although its actual rotation period is two-thirds of its orbital period. Alternatively, the physiology of the eye may lead to mistakes. For example, the *teardrop effect* during *transits* of Mercury or Venus arises from cross-links between the rods and cones in the eye's retina. See also: *irradiation.*

fan beam

A *main lobe* for a *radio telescope* that is narrow in one axis and broad in the other. See also: *pencil beam.*

F

Faraday rotation

A rotation of the direction of *polarization* of *linearly polarized electromagnetic radiation* as it passes through a magnetized plasma. It is of astronomical significance only in the *radio region.*

far field response

The response of an *antenna* for objects sufficiently far from the antenna that the antenna's dimensions are negligible. See also: *near field response.*

fast Fourier transform

A computer algorithm for calculating *Fourier transforms* and inverse transforms very quickly and efficiently. The number of data points must, however, be a power of 2, and so actual data sets usually have to be bulked out by adding zeros.

fat zero

The practice of exposing a *CCD* detector to a low-level uniform background illumination. It reduces the effects of read-out *noise* on the faintest parts of the image, and is removed later during *image processing.*

feed

An *antenna* placed at the focus of a dish or other type of *radio telescope* in order to convert the incoming radio waves to an electrical signal. The feed is often a *half-wave dipole* or *Yagi* antenna.

Fellgett advantage

A synonym for *multiplex advantage.*

FFT

See *fast Fourier transform.*

fiber optics

Thin glass fibers that are able to pipe light along their lengths through multiple internal reflections. They are being increasingly used to connect telescopes to *spectroscopes*, especially for *multi-object spectroscopy.*

field curvature

A fault of images in which the focal surface (i.e. the surface wherein lie the most sharply focused points of the image) is curved not flat. In order to produce sharp images over the whole field of view, the detector therefore has to be curved to match the shape of the focal surface, as in the *Schmidt camera.* See also: *aberration (optical).*

field flattener
A *correcting lens* designed to compensate for *field curvature*, and to produce a plane focal surface.

field lens
The first lens of an *eyepiece* that gathers the light from the *objective* and which is furthest from the observer's eye. See also: *eye lens*.

field of view
The angular size of the maximum object completely encompassed by an optical system. *Eyepieces* have fields of view ranging from 30° to 80°. The field of view of a telescope is the field of view of the eyepiece divided by the *magnification*. A telescope used at ×200 with an eyepiece whose field of view is 50° will thus have a field of view of just 15′.

F

field pattern
See *polar diagram*.

field star
A star (or other object) in the same *field of view* as a galaxy, star cluster, etc. but not a part of that galaxy or cluster. If closer to us than the main object, it is a foreground star, if further away, a background star.

field stop
See *stop*.

figure
The shape of an optical surface.

figuring
The process during the production of a lens or mirror whereby the precise shape required is imparted to the surface. The depth of material removed is only a few microns, but the process makes all the difference between a poor quality instrument and one that is *diffraction limited*. Figuring is undertaken by continued *polishing* of specific areas of the optical surface in order to reduce high points to the required profile. It must be undertaken in conjunction with optical tests of the accuracy of the surface such as the *Foucault test*, the *Hartmann test* or the *Ronchi test*, in order to monitor the effects of the polishing. See also: *grinding*.

filar micrometer
See *bi-filar micrometer*.

filled aperture
See *aperture synthesis*.

filter
1 An optical device that absorbs or reflects some wavelengths whilst allowing others to pass through. It may be a cut-off filter that absorbs below some wavelength and transmits above it (or vice versa), or a *band-pass filter* that transmits only over a certain range (the *band width*) of wavelengths. *Cut-off* and *wide band filters* are usually based upon chemical dyes that absorb the required wavelengths and are incorporated into optically flat glass plates. *Narrow band filters* are usually based upon *Fabry–Perot etalons*. See also: *H–α filter; light pollution rejection filter; optical density.*

Color filters. (Image reproduced by courtesy of Meade Instrument Corp.)

2 An electronic device that transmits signals within a certain frequency range and attenuates others. A low pass filter allows through signals below a certain frequency, while a high pass filter allows through signals above a certain frequency. These filters are the equivalent of the optical cut-off filter. A band pass filter is the equivalent of the optical band pass filter and allows through only a certain range of frequencies. A notch filter is the inverse of a band pass filter and attenuates signals over a certain range of frequencies. All electronic equipment has a limited range of frequency response, and therefore acts as a filter whatever may be its primary purpose. The term is therefore reserved for devices whose principal purpose is to limit the frequency of the signal.
3 An *image processing* technique, sometimes known as a spatial filter. The filter changes the intensity of a given *pixel* by combining it with the intensities of surrounding pixels. The action of the filter may be represented as a 3 × 3, 5 × 5, 7 × 7, etc. matrix. The elements of the matrix are the weightings given to each pixel intensity when they are combined and used to replace the original intensity of the central pixel. Filters can act to *smooth* or *sharpen* an image, or enhance edges, etc.

+1/9	+1/9	+1/9
+1/9	+1/9	+1/9
+1/9	+1/9	+1/9

(a) A smoothing filter

+1/16	+1/16	+1/16
+1/16	+1/2	+1/16
+1/16	+1/16	+1/16

(b) A weaker smoothing filter

+1/25	+1/25	+1/25	+1/25	+1/25
+1/25	+1/25	+1/25	+1/25	+1/25
+1/25	+1/25	+1/25	+1/25	+1/25
+1/25	+1/25	+1/25	+1/25	+1/25
+1/25	+1/25	+1/25	+1/25	+1/25

(c) A wider smoothing filter

-1	-1	-1
-1	+9	-1
-1	-1	-1

(d) A sharpening filter

-0.5	-0.5	-0.5
-0.5	+5	-0.5
-0.5	-0.5	-0.5

(e) A weaker sharpening filter

-1	-1	-1
-1	+8	-1
-1	-1	-1

(f) An edge enhancement filter

Image processing filters:
a smoothing;
b weaker smoothing;
c broader smoothing;
d sharpening;
e weaker sharpening;
f edge enhancement.

filter – H-α
See *H-α filter.*

filter-gram
See *spectroheliogram.*

finder chart
A map or image of a part of the sky containing an object of interest. The finder chart is used when observing to enable that object to be distinguished from the other objects within the telescope's *field of view*. When preparing a finder chart it is best to arrange its orientation and scale to be similar to that seen through the telescope. Otherwise it is very easy to make mistaken identifications.

finder telescope
A small telescope with a wide *field of view* that is attached to and aligned with a larger telescope. It usually has a *cross-wire eyepiece*. It is used for *finding* objects in the sky. See also: *red-dot finder.*

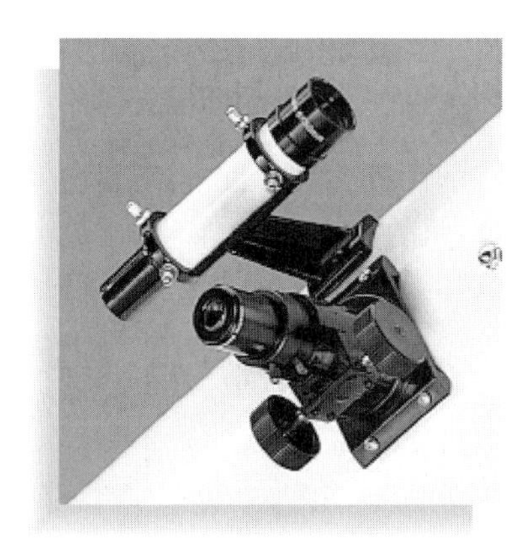

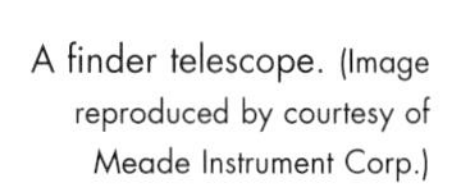

A finder telescope. (Image reproduced by courtesy of Meade Instrument Corp.)

finding

The process of directing a telescope towards an object in the sky so that it may be seen through the *eyepiece* or centered on a *detector*. There are various ways of finding objects:

- With accurate *setting circles* or a computer-controled instrument, the telescope may be moved directly to the tabulated position of the object and seen through the main instrument.
- For moderately bright objects, the telescope may be pointed in approximately the correct direction (*setting*, it is useful to be familiar with the brighter *constellations* for this purpose), and then identified in the *finder telescope*. A *zero power finder* may also be used at this stage. The telescope is then moved to center the object on the finder telescope's *cross-wires*, and it will then be visible in the main instrument.
- By star hopping. This method may be used for objects that are too faint to see in the finder telescope. A good *finder chart* is needed which shows the stars visible in the main telescope along the path between the object of interest and a nearby brighter star. The brighter star is first found, and then while looking through the main telescope, it is moved successively to stars along that path until the object of interest is reached.

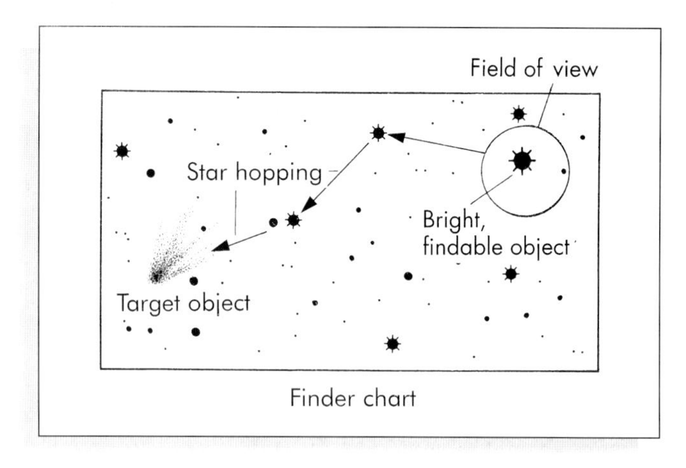

Star hopping.

finesse

A measure of the *spectral resolution* of a *Fabry–Perot etalon*. It is related to the normal spectral resolution, R, by:

$$Finesse = \frac{\lambda R}{2t},$$

where λ is the wavelength, and t the separation of the partially reflecting mirrors of the etalon. It is also related to the reflectivity, r, of those mirrors by:

$$Finesse = \frac{\pi\sqrt{r}}{1-r}.$$

First Point of Aries

The point in the sky where the *ecliptic* and the *celestial equator* cross and where the Sun's annual motion takes it from south to north of the equator. It is also called the *vernal* or *spring equinox*, and the Sun passes through it on or about 21 March every year. It is the reference point for *right ascension* and *celestial longitude* and its *hour angle* gives the local *sidereal time*. Currently it is to be found in Pisces near the border with Cetus and Aquarius (it may be found in the sky by taking a line from β Peg through the center of the line between γ Peg and α Peg and projecting for the same distance again), having moved from its original position in Aries due to *precession*.

first quarter

The phase of the waxing Moon when its disk is exactly half illuminated. See also: *Half Moon; last quarter; dichotomy.*

FITS

See *electronic image format*.

fixing
A chemical process that dissolves away the unwanted silver bromide crystals from a *photographic emulsion* after it has been *developed.*

FK5
See *fundamental catalog.*

Flamsteed number
A system of *stellar nomenclature* deriving from John Flamsteed's (1646–1719) catalog of 1725. The stars are numbered within each *constellation* in order of increasing *right ascension.* For example, 57 UMa, 88 Leo or 15 Pic. See also: *Bayer system.*

F

flash spectrum
The *spectrum* of the solar chromosphere. So called because it flashes into view for just a few seconds between the start of a total eclipse, and when the Moon covers the chromosphere (plus a similar few seconds just prior to the end of totality). It is an *emission-line spectrum,* and if the slit of a conventional *spectroscope* is dispensed with, the flash spectrum will consist of a series of images of the chromosphere in each of its emission lines. (CAUTION – always take appropriate precautions when observing the Sun or damage to the eye and/or telescope can result).

The flash spectrum of the solar chromosphere. (Image reproduced by courtesy of the Royal Astronomical Society.)

flat
A plane mirror. Such as the *secondary mirror* of a *Newtonian telescope.* See also: *optical flat.*

flat fielding
An *image processing* technique applied to *CCD* images. The CCD's response will vary slightly over the image. This is corrected by obtaining an image of a uniformly illuminated object (the flat field), and dividing the main image by it. The flat field will be more intense in those areas where the CCD has a higher sensitivity, and so this process will proportionally reduce the too bright parts of the main image. Flat fields may be obtained by imaging a sheet of white card on the inside of the telescope's dome, or the twilit clear sky, etc.

flattening, polar
A synonym for *oblateness.*

flexible image transport system
See *electronic image format.*

flexure
The bending of the structure of an instrument under gravitational or other loadings. For optical instruments flexure may result in the optical components being displaced from their correct positions, and so severe *aberrations* introduced into the image. Small instruments may be made sufficiently strongly that flexure effects are negligi-

ble. Larger instruments, however, especially large *optical* and *radio telescopes*, have to be designed to correct the effects of flexure (see *Serrurier truss* and *homological transform designs*).

flicker noise

A *noise* source, whose mechanism is not well understood, that occurs in signals which are varying with time. The signal variations may be intrinsic or arise through the *chopping* of the signal. Flicker noise tends to reduce in significance as the frequency of the variations increases.

flint glass

A type of glass with a *refractive index* around 1.6 and a *constringence* of 30 to 35 that is often used to form the diverging component of an *achromatic lens*. See also: *crown glass*.

flip mirror

A mirror incorporated into an ancillary instrument for a telescope, such as a *photometer*, that may be moved (flipped) into the optical path to reflect light to a *finding system*, and then moved out of the optical path to allow the measurements to take place.

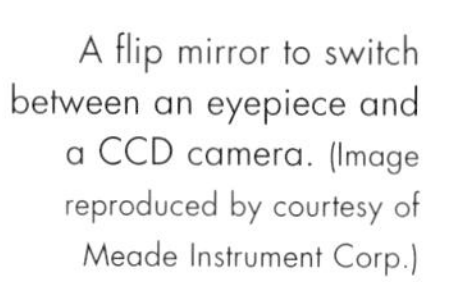

A flip mirror to switch between an eyepiece and a CCD camera. (Image reproduced by courtesy of Meade Instrument Corp.)

flux

The rate at which energy (or occasionally, number of particles, magnetic field, etc.) flows through a surface. It is most usually used in respect of the emission of *electromagnetic radiation* from a surface, when its SI units are W m^{-2}.

flux collector

A telescope whose primary purpose is to gather energy, not to produce high quality images, also known as a light bucket. It will therefore generally be cheaper to construct than a normal telescope since the optical surfaces do not need to be of high quality.

flux density

In radio astronomy, the power received from a source per unit wavelength or frequency interval. Its SI units are W m^{-2} Hz^{-1}, but a unit called the *jansky* (Jy) is more commonly used, 1 Jy = 10^{-26} W m^{-2} Hz^{-1}.

flux unit

A synonym for a *jansky*.

f-number

See *f-ratio*.

focal length

The distance between a lens or mirror and the image produced from a parallel beam of incident radiation. For *converging lenses* and concave mirrors, the image will be *real*, for *diverging lenses* and convex mirrors it will be a *virtual image*.

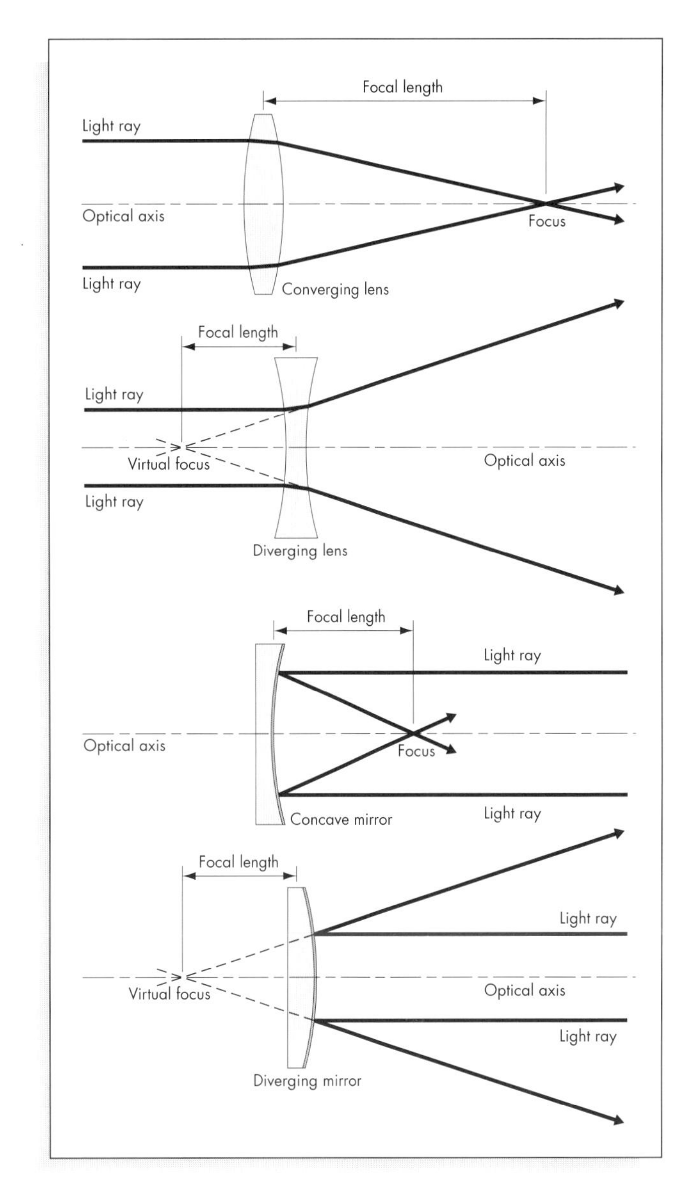

Focal length.

focal plane

A flat surface, orthogonal to the *optical axis*, wherein lie the most sharply focused points of the image. In some instruments, such as the *Schmidt camera*, the focal surface is curved not flat – see *field curvature*.

focal point

See *focus*.

focal ratio

See *f-ratio*.

focal reducer

An optical component, usually a *positive achromatic lens* that reduces the *effective focal length* of a telescope. It is positioned just inside the *focal point* of the telescope *objective*. Its purpose is to reduce the *image scale* and to increase the *field of view* of the telescope.

focal surface

See *focal plane*.

focus

1 A point at which a beam of radiation is caused to converge by an optical system (*real image*), or from which the light beam diverges (*virtual image*). For a parallel incident beam of radiation, the focus is at a distance called the *focal length* away from the lens or mirror. In a telescope used visually, the image is seen to be in focus when the light rays emerge as a parallel beam. The actual focus and image is then produced on the retina of the eye by the eye's lens.
2 One of two points inside an *ellipse*. For planetary orbits, the Sun occupies one of these points. The name arises because a source at one focus will be imaged at the other.

focusing

The process of adjusting an optical system to make the image as sharp as possible. For a telescope used visually this is accomplished by moving the *eyepiece* along the *optical axis*. When imaging, the *detector* may be moved in a similar fashion, or a subsidiary optical system may be used. In either case, for *Cassegrain* and telescopes of similar design, focusing may also be undertaken by moving the *secondary mirror* along the optical axis. This method of focusing, however, should be used with care since displacing the secondary mirror from its optimum position introduces *aberrations* into the image. See also: *Hartmann mask*.

focuser

A device at the focal point of a telescope into which *eyepieces*, *cameras*, and other ancillary instrumentation may be mounted and their positions adjusted until the image is focused. A focuser may be a simple *draw tube* that is pushed in and out to focus, or of a more elaborate construction that enables focusing to be made more smoothly and precisely. Two common means of accomplishing this are by the helical focuser which screws in and out through the use of a large diameter thread on the outside of the mounting, and the rack and pinion in which a small gear (the pinion) drives a linear toothed bar (the rack) attached to the moving part (slider) of the focuser. Cheaper versions of the latter use a friction wheel acting directly onto the side of the slider. Both of these designs may be motorized to produce an electrical focuser. A zero-image-shift focuser is particularly useful for focusing onto small CCD chips since the image does not move laterally when the focusing direction is changed. The addition of a scale or position read-out onto the focuser is also very useful since the focal positions of particular instruments, eyepieces, etc. can be noted for future use. Since the focuser may alter with temperature, some manufacturers now produce temperature-controled versions. See also: *eyepiece mount*.

A temperature-controled electric focuser. (Image reproduced by courtesy of Optec Inc.)

fog

The background *noise* on a *photographic emulsion*. It comprises grains that have been *developed* but which were not exposed to light.

following

See *trailing edge*.

forbidden line

A *spectrum line* that arises from a *transition* not allowed under the normal rules. It may arise through a breakdown of the normal rules caused by electron interactions in complex atoms, or through special physical conditions such as those in interstellar nebulae. A forbidden line is denoted by the use of square brackets, for example; [O II 731.99] (see *ionization* for further details of this notation).

The degree to which a line is forbidden can range from a 1% probability of occurrence compared with a normal line to 10^{-20} % or less. At the 1% level, forbidden lines can occur as weak absorption lines in stellar spectra. At lower levels of probability, forbidden lines are usually emission lines. In interstellar nebulas the strongest lines are forbidden lines due to oxygen and nitrogen, [O III 495.9], [O III 500.7], [O II 372.6], [O II 372.9] and [N III 386.9] and these arise because the higher *energy level* of the transition becomes over-populated through collisional *excitations*. The 21-cm line of atomic hydrogen is an example of a forbidden line at the lowest level of probability of occurrence that is none-the-less observed because of the vast amount of hydrogen in interstellar space. See also: *intercombination line.*

foreground star

See *field star.*

fork mounting

A design of *equatorial mounting* for a telescope in which the telescope swings in *declination* between a pair of arms (or fork) projecting from the top of the *polar axis.* The design is often used as the mounting for small *Schmidt–Cassegrain telescopes.*

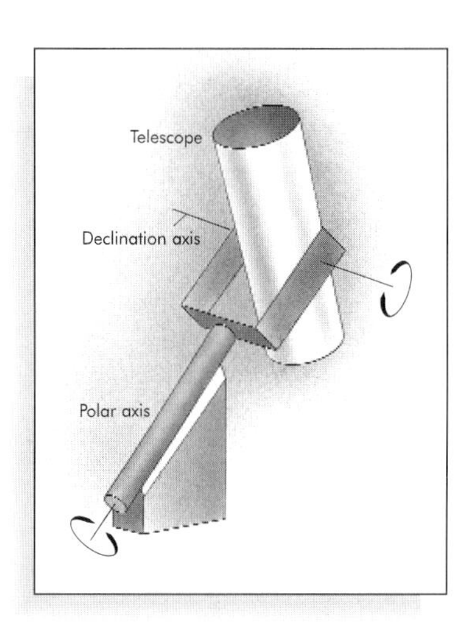

The fork telescope mounting.

format–image

See *electronic image format.*

forming gas

Nitrogen with the addition of between 2% and 10% hydrogen, used in some methods of *hypersensitizing* photographic emulsions.

Foucault test

A test for the shape of the surface of a concave mirror, also known as the knife-edge test, and devised by Leon Foucault (1819–68). An illuminated slit is placed at the radius of curvature (twice the *focal length*) of the mirror. The reflected beam is focused next to the light source and is observed directly by the eye. When the reflected beam is intercepted by a sharp edge moved gradually across it and near to the focus, a series of shadows appear on the mirror. The nature and depth of the shadows are used to determine the exact shape of the mirror's surface to an accuracy of a tenth of the wavelength of light or better. See also: *figuring; Hartmann test; Ronchi test.*

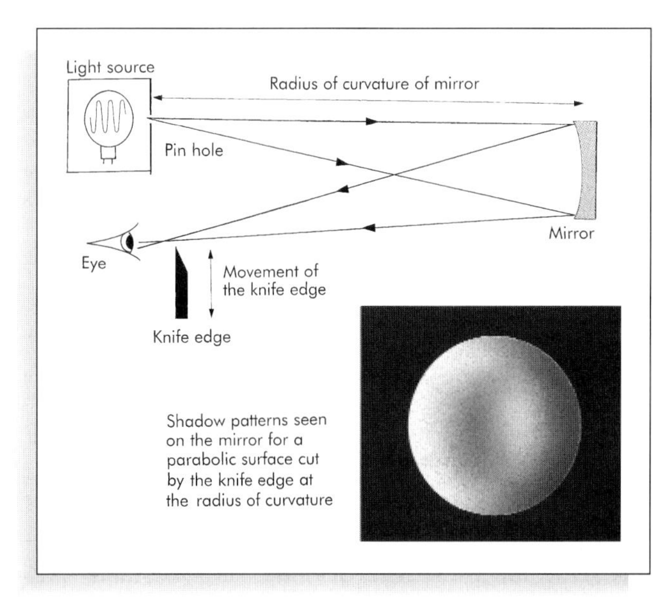

The Foucault test for mirrors.

Fourier analysis

A mathematical operation whereby a periodic but complex function is broken down into a sum of an infinite series of simple sine and cosine terms of differing frequencies, called the Fourier series of the function. The process is perhaps more familiar as the note from a musical instrument being split up into its fundamental and higher harmonic frequencies. If the function is denoted by f(x), and has a periodicity of 2π, then its Fourier series is given by:

$$f(x)=\frac{a_0}{2}+\sum_{n=1}^{\infty}(a_n \cos nx+b_n \sin nx),$$

where

F

$$a_n=\frac{1}{\pi}\int_{-\pi}^{\pi} f(x)\cos nx\,dx \quad (\mathrm{n}=0-\infty)$$

and

$$b_n=\frac{1}{\pi}\int_{-\pi}^{\pi} f(x)\sin nx\,dx \quad (\mathrm{n}=1-\infty).$$

For example a square wave may be represented by the series:

$$f(x)=\sin x+\frac{1}{3}\sin 3x+\frac{1}{5}\sin 5x+\frac{1}{7}\sin 7x+\dots.$$

See also: *string method.*

Fourier deconvolution

See *deconvolution.*

Fourier series

See *Fourier analysis.*

Fourier transform

A mathematical operation whereby a non-periodic function is represented as a function of its component frequencies (cf *Fourier analysis*). The Fourier transform of a function, f(x), denoted by F(t), is given by:

$$F(t)=\int_{-\infty}^{\infty} f(x)e^{-2\pi ixt}\,dx,$$

while the inverse Fourier transform reverses the process:

$$f(x)=\int_{-\infty}^{\infty} F(t)e^{2\pi ixt}\,dt.$$

The power spectrum is a plot of the power in the original signal at individual frequencies, and is obtained by plotting the magnitude of the Fourier transform against frequency. Fourier transforms are used in obtaining images from *aperture synthesis systems*, obtaining spectra from *Fourier transform spectroscopes*, in determining the periodicities present in complex data, and in *deconvolution*. See also: *fast Fourier transform; string method.*

Fourier transform spectroscope

A device for producing a *spectrum* based upon the *interferometer* used in the Michelson–Morley experiment, also known as a Michelson spectroscope. The light from the telescope is split into two orthogonal beams by a *beam splitter*. The beams are then reflected back to the beam splitter and recombined. Path differences between the beams leads to an interference pattern at the device's output that is monitored by a point *detector*. One of the two mirrors reflecting back beams towards the beam splitter is then slowly moved along the length of the beam. The resulting changes in path differences between the beams leads to changes in the interference pattern and so in the output from the detector. The output from the detector is a section of the *Fourier transform* of the spectrum of the incoming radiation. Taking the inverse Fourier transform of the output therefore provides the required spectrum. See also: *multiplex advantage.*

fovea centralis

See *retina.*

FPA

See *First Point of Aries.*

fractional estimate

A method of determining the magnitude of a star from a visual estimate. Two *standard stars* whose brightnesses bracket that of the star of interest and whose magnitudes are known are required in the same *field of view* as the star of interest. The brightness of the star of interest is estimated as the number of tenths of the brightness interval between the standard stars that it is brighter than the fainter standard. The estimate is then recorded in the form (A n B), where A is the fainter standard, B, the brighter standard and n the number of tenths difference from A for the star of interest. The technique is mostly used to monitor variable stars. The same standards can then be used on every observing occasion. With practice, estimates correct to 0.1^m can be reached.

f-ratio

The ratio of the *focal length* of a lens or mirror to its diameter or *aperture*. For camera lenses it is also known as the speed of the lens, and may be adjusted by means of an internal diaphragm within the lens. Telescope f-ratios range from 2 or 3 for the *primary mirrors* of modern *reflectors*, through 8 or 10 for *Newtonian telescopes* and the *Cassegrain* foci of telescopes, 10 to 15 for *refractors*, and 20 or more for *Coudé* foci.

Fraunhofer interference

The *interference* pattern of *electromagnetic radiation* at a large distance from the interaction region (see *far field response*), or when the light beams have been brought to a focus by a lens or mirror.

Fraunhofer lens

An *achromatic lens* with the two components having unequal radii, and an air space. This design generally gives the highest levels of correction but is the most difficult to construct and to keep in alignment. See also: *Littrow condition; Clairault condition.*

Fraunhofer lines

The most prominent *spectrum lines* in the visible solar spectrum, first labelled in 1814 by Joseph von Fraunhofer (1787–1826). They are:

Line	Wavelength (nm)	Origin
A	759.4	Molecular Oxygen (Terrestrial)
B	686.7	Molecular Oxygen (Terrestrial)
C	656.3	Hydrogen (H-α)
D_1	589.6	Sodium
D_2	589.0	Sodium
E1	527.0	Iron
Eb	518.3–516.7	Magnesium triplet
F	486.1	Hydrogen (H-β)
G	430.8	Iron plus CH molecule (now, G′ is often used for H-γ at 434.0)
H	396.8	Calcium

K	393.3	Calcium
L	382.0	Iron)
M	373.5	Iron) Added after Fraunhofer's time
N	358.1	Iron).

Fraunhofer mounting
A synonym for a *German mounting.*

Fred Lawrence Whipple observatory
An optical observatory, originally named the Mount Hopkins observatory, sited on Mount Hopkins, Arizona at an altitude of 2300 m. It houses the *multi-mirror telescope*, amongst several others.

F

free-bound transition
See *recombination.*

free-free transition
Radiation emitted (or occasionally absorbed) as a loose electron goes past an ion, also known as bremsstrahlung radiation (from the German for 'braking'). The name arises because the electron under goes a transition from one *energy level* that is 'free' of the ion to another that is 'free' of the ion. See also: *synchrotron radiation.*

free spectral range
The separation in wavelength terms of two wavelengths that are physically superimposed but from adjacent *spectral orders* for *spectra* produced by a *diffraction grating* or *etalon* based *spectroscope.* It is given approximately by the wavelength divided by the spectral order.

frequency
The number of maxima (or minima) of a wave motion passing a given point per second, usually symbolized by ν or f. See also: *wavelength; period.*

frequency analysis
See *Fourier transform.*

Fresnel interference
The *interference* pattern of *electromagnetic radiation* at a small distance from the interaction region (see *near field response*).

Fresnel lens
A lens design of low optical quality but in which the thickness and weight of the lens is greatly reduced. The surface of the lens consists of a series of narrow annuluses each of which has the same shape as the corresponding surface of a normal lens, but with a reduced thickness. A cross section through the lens thus has a 'stair-case' shape. The most familiar use of a Fresnel lens is probably within over-head projectors as used in lecture theatres.

Fried's coherence length
The maximum diameter of a telescope before atmospheric turbulence starts to limit its resolution. It is given by:

$$D \approx 0.114\left(\frac{\lambda \cos z}{550}\right) \quad m,$$

where λ is the wavelength in nanometers, and z is the zenith distance. See also: *Rayleigh resolution; seeing; scintillation.*

fringe pattern

See *Airy disk* and *interference.*

FTS

See *Fourier transform spectroscope.*

full-aperture filter

See *solar filter.*

F

Full Moon

The phase of the Moon when it is in *opposition* to the Sun, and so we see its disk fully illuminated. See also: *New Moon; Half Moon.*

full width at half maximum

A measure of the width of an object or function, often abbreviated to *FWHM.* It is usually used when the actual edges of the object are imprecise or difficult to determine (i.e. they may be 'fuzzy' or lost in *noise*). For example, an intensity tracing across the head of a comet or over a spectrum line. The width is then measured at half the maximum height (or depth) of the function, where the edges are still reasonably well defined. The half-width at half maximum (HWHM) is also used and is just half the FWHM.

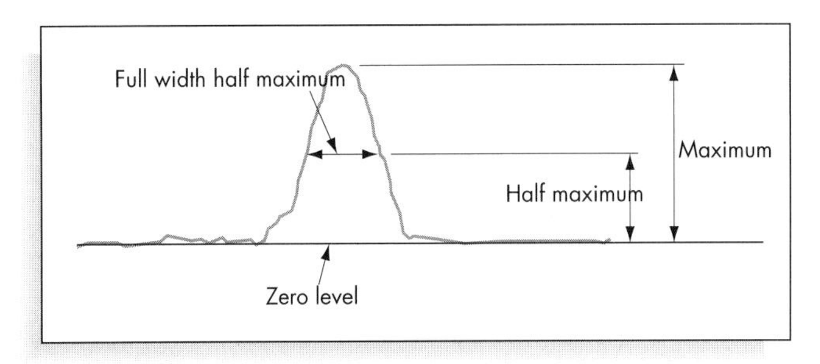

Full width, half maximum (FWHM).

fundamental catalog

A catalog listing very precise positions and proper motions of stars and which provides a reference frame for other star positions. The *International Astronomical Union* has adopted the fifth of a series of German catalogs: Fundamentalkatalog, known as FK5, as the current standard.

fundamental frequency

The lowest frequency of an oscillating system.

FWHM

See *full width half maximum.*

G

gain

The ratio of the maximum signal received from an *antenna* when acting as a transmitter to the signal that would be received if the same total energy were to be broadcast equally in all directions. For a *radio receiver* it is in effect the amplifying factor obtained by using the antenna. See also: *polar diagram; directivity; effective aperture.*

galactic latitude

A coordinate system, along with *galactic longitude*, for giving the positions of objects in the sky. The galactic latitude (symbol, b) is the angle, in degrees, up or down from the galactic equator, going from 0° to ±90°. When the system was initially set up an incorrect direction for the center of the galaxy was used. Positions based on that direction are sometimes designated as b^{I}, with the correct positions designated b^{II}. Galactic latitude is related to the *RA* (α) and *Dec* (δ), of the object by:

$$\sin b^{II} = \sin \delta \cos 62.6° - \cos \delta \sin(\alpha - 282.25°)\sin 62.6°$$

Note that the RA must be converted to degrees before inserting into this formula. See also: *declination; right ascension; celestial sphere.*

galactic longitude

A coordinate system, along with *galactic latitude*, for giving thc positions of objects in the sky. The galactic longitude (symbol, l) is the angle, in degrees, around the galactic equator from the direction to the center of the galaxy (at RA 17h 42m 24s, Dec −28° 55′, in Sagittarius, the actual center is now known to be about 4′ away from this point, but this position is still used for the galactic coordinates), going from 0° to 360°. When the system was initially set up an incorrect direction for the center of the galaxy was used. Positions based on that direction are sometimes designated as l^{I}, with the correct positions designated l^{II}. Galactic longitude is related to the *RA* (α) and *Dec* (δ), of the object by:

$$\cos (l^{II} - 33°) = \frac{\cos \delta \ \cos(\alpha - 282.25°)}{\sqrt{1 - (\sin \delta \ \cos 62.6° - \ \cos \delta \ \sin (\alpha - 282.25°) \ \sin 62.6°)^2}}.$$

Note that the RA must be converted to degrees before inserting into this formula. See also: *declination; right ascension; celestial sphere.*

galactic year

The orbital period of the Sun around the galaxy, about 225 million years.

Galilean telescope

A *refracting telescope* that uses a *diverging lens* as the *eyepiece*. First designed and used by Galileo Galilei (1564–1642), it provides *erect images*, the *field of view* and *eye* relief, however, are small. Apart from being used for opera glasses, it has now been completely superseded by other telescope designs.

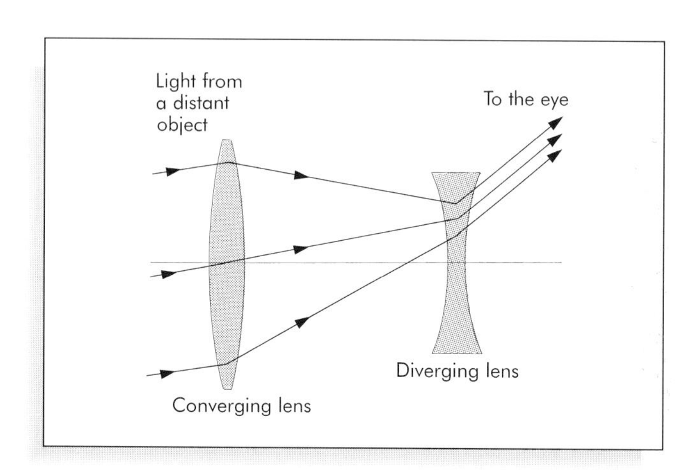

The light paths in a Galilean telescope.

GALLEX neutrino detector.

A *neutrino detector*, based upon detecting the conversion by a neutrino of gallium-71 to radioactive germanium-71. It uses 30 tonnes of liquid gallium chloride and is buried in the Grand Sasso tunnel in Italy.

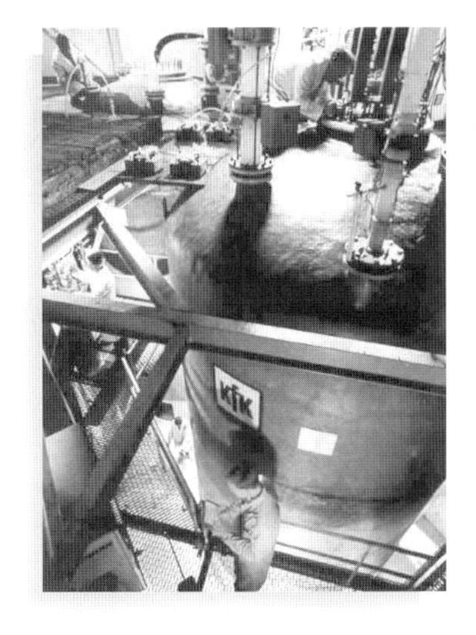

The tank containing 30 tonnes of gallium chloride for the GALLEX neutrino detector. (Image reproduced by courtesy of the GALLEX collaboration, T. Kirsten, Heidelberg.)

gamma

A measure of the contrast of a *photographic emulsion*. It is the tangent of the slope of the linear portion of the *characteristic curve*. See also: *contrast index*.

gamma ray

The shortest *wavelength*, highest *frequency*, form of *electromagnetic radiation*. The wavelength is 0.01 nm or less and the frequency 3×10^{19} Hz or more. In energy terms, gamma ray photons have energies of 10^5 eV or more.

gamma ray detector

A detector for *gamma rays*. Since the Earth's atmosphere is opaque to gamma ray radiation, gamma ray detectors have to be flown on rockets or spacecraft. They can also normally be used to detect X-rays and cosmic rays, and include *proportional counters*, *scintillation detectors*, *solid state detectors*, *nuclear emulsions* and *spark detectors*.

gamma ray telescope

No instruments exist to permit the direct imaging of *gamma ray* sources. The incoming direction of a gamma ray, however, may be found from *spark detectors*. Other forms of imaging require two or more detectors to be used together. Simultaneous detections must then arise from gamma rays travelling along the line between the detectors (*coincidence detection*). Alternatively images may be built up by *scanning* and using a *collimator* or *shielding* to restrict the *field of view* of the detector. See also: *coded array mask; modulation collimator; X-ray detectors*.

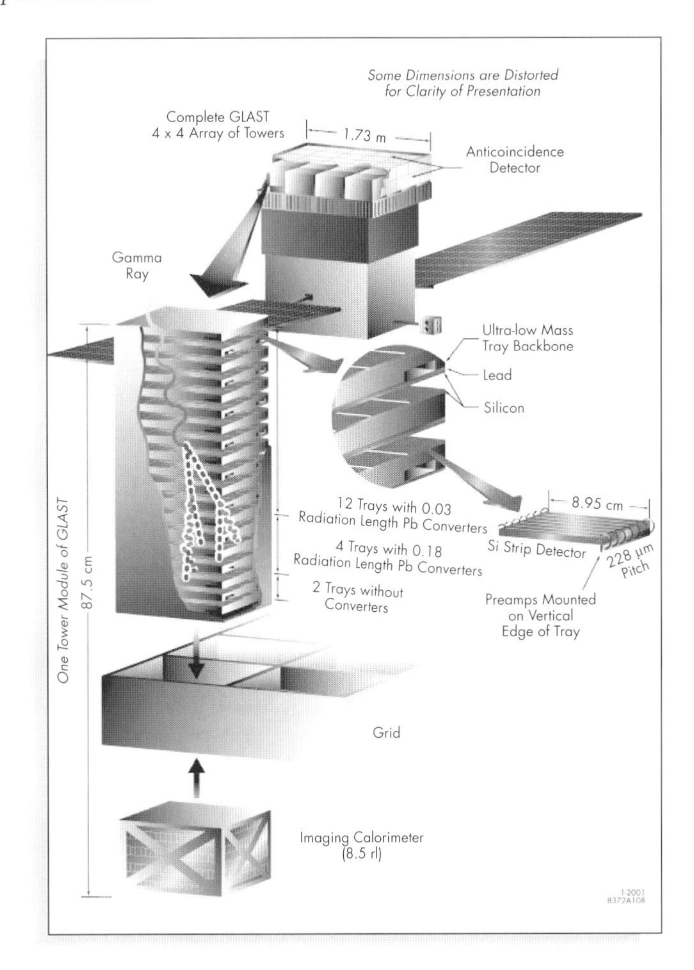

The proposed Gamma Ray Large Area Space Telescope (GLAST). (Image reproduced by courtesy of NASA/SLAC.)

gas scintillation proportional counter

See *proportional counter.*

Gaussian distribution

A bell-shaped curve that represents the distribution of multiple measurements of the same quantity, where the differences between the measurements are just the result of random errors. It is also called the normal distribution. The *standard deviation* is half the width of the Gaussian curve at a height $e^{-0.5}$ (= 61%) of its peak. The mathematical form of the curve is:

$$P(x) = \frac{1}{\sigma\sqrt{2\pi}} e^{-\frac{(x-X)^2}{2\sigma^2}},$$

where $P(x)$ is the probability of a measurement, x, being obtained, X is the mean value of the measurements and s is the standard deviation.

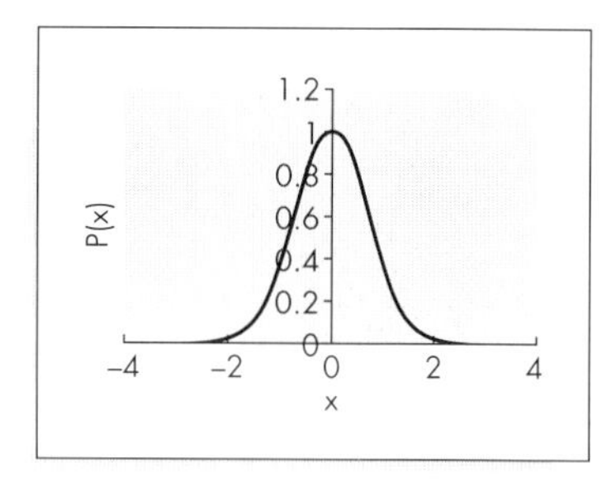

The Gaussian (or normal) distribution.

G

GCVS

See *general catalog of variable stars.*

Gemini telescopes

See panel (page 105).

general catalog of variable stars

A catalog of variable stars and their properties containing nearly 30,000 entries and last published, in Russia, in 1987.

GEO 600

See *gravity wave detector.*

geocentric coordinates

Coordinates (e.g. *right ascension* and *declination*) as they would be seen from the center of the Earth. See also: *heliocentric coordinates; topocentric coordinates.*

geocentric parallax

See *diurnal parallax.*

geodesy

The study of the shape and size of a planet, especially the Earth.

germanium detector

See *solid-state detector.*

German mounting

A design of *equatorial mounting* for a telescope in which the *declination axis* is mounted on the top end of the *polar axis.* The telescope is mounted on one side of the declination axis with a counter weight on the other side.

A German mounting for a small telescope. (Image reproduced by courtesy of Meade Instrument Corp.)

Gemini telescopes

Two 8.1m optical telescopes using monolithic thin mirrors held in shape by *active supports*. One instrument (Gemini North) is on Mauna Kea, Hawaii, altitude 4200 m, the other (Gemini South) on Cerro Pachón, Chile, altitude 2700 m. See homepage at http://www.gemini.edu/.

A panorama inside the Gemini North dome. (Image reproduced by courtesy of Gemini Observatory.)

The Gemini North telescope. (Image reproduced by courtesy of the Gemini observatory.)

One of the Gemini mirrors during production. (Image reproduced by courtesy of Gemini Observatory.)

A photo-montage showing the Gemini North dome with the stars visible in early Spring at Mauna Kea behind it. (Image reproduced by courtesy of Gemini Observatory.)

G

ghost

An unwanted or false feature within an image. Ghosts often arise through multiple reflections between and within optical components. Since such reflections are from plain glass, the intensity of a ghost is much less than that of the main image. Confusion may arise, however, when the ghost is from a bright object outside the field of view. Ghosts can sometimes be identified through their being out of focus, or if the telescope is moved, the ghost may move in a different direction from the main image.

In *spectroscopy*, however, ghosts may arise as lines from higher or lower order spectra or as the result of periodic errors in the rulings of the *diffraction grating* (known as Lyman and Rowland ghosts), or from spectral orders from behind the grating reappearing in front of it (Wood's anomalies) and these are much more difficult to identify.

G

giant meter-wave telescope

A radio *interferometer* with thirty 45m dishes located at Khodad, Maharashtra, Western India.

gibbous phase

The shape of the Moon, Mercury or Venus, when more than half of the illuminated surface is visible. The outer planets are normally also gibbous in shape (except at *opposition*, when they are 'full' – i.e. the whole of the illuminated surface is visible), though the effect is usually only observable for Mars. See also: *crescent phase.*

GIF

See *electronic image format.*

glancing incidence telescope

See *grazing incidence telescope.*

Glan–Thompson prism

A *polarizer* that produces a single linearly polarized beam. It consists of two right-angled prisms with parallel optical axes, cemented together along their long sides. It is similar to the *Nicol prism*, except that in the latter the prisms are cemented along their bases.

glitch

A sudden disturbance in an otherwise regular series of events. It is applied especially to the occasional abrupt increase in pulse frequency observed for some pulsars.

global oscillation network group

A group of six solar observatories distributed around the world in order to give near-continuous coverage of the Sun, and dedicated to *helioseismology*, and with the acronym GONG.

GMT

Abbreviation of Greenwich Mean Time – see *universal time.*

gnomon

The upright part of a *sundial* that casts a shadow onto the time scale. See also: *style.*

Goddard space flight center

A *NASA* organization at Greenbelt, Maryland, concerned with space research.

Golay cell

A *detector* for *infrared radiation* based on the heating effect of the radiation on gas in an enclosed bulb. Now completely superseded by more sensitive detectors.

GONG

See *global oscillation network group.*

goto capability

A feature of many modern computer-controled telescopes of all sizes that allows the observer to key in the object's name or a simple code, and the telescope then automatically sets onto that object. The computer has a data base of the positions of thousands of objects in the sky and the orbital elements for moving objects like the planets and asteroids. It then calculates their positions in the sky for the time and place of the observation, usually also correcting for precession and refraction, etc.

G

Goto Optical

A commercial firm manufacturing planetaria.

Gran telescopio Canarias

A 10m *Ritchey–Chrétien* optical telescope using 36 hexagonal segments held in position by *active supports* to form the primary mirror. Sited on Las Palmas, Canary islands and due to open in 2002.

graticule

A pattern of fine lines at the focus of an eyepiece that enables precision timing or position measurements to be made, also called a reticle. The lines may be etched onto an optically flat piece of glass, or be thin threads stretched across the eyepiece. The arrangement of the lines may be as a simple *cross-wire*, several parallel lines, a grid, concentric circles, etc. See also: *astrometric eyepiece.*

grating spectroscope

A *spectroscope* based upon a *diffraction grating.*

gravity wave detector

A detector for gravity waves. Gravity waves are oscillations of the space–time continuum (gravity field) produced when a mass is accelerated. They are predicted to occur by general relativity and most other theories of gravity, but have yet to be detected unequivocally. Changes to the orbit of the binary pulsar (PSR 1913 +16) are, however, consistent with their existence. Gravity wave detectors are based upon the tidal distortion, or strain, that a gravity wave is expected to produce in material objects as it passes through them. This distortion is very small. For the strongest predicted waves it is expected to be about 10^{-21} m (0.00000000001 of the diameter of a hydrogen atom) in an object 1m long. The earliest detector was produced by Joseph Weber in the 1960s and consisted of a massive aluminum cylinder that would vibrate if stimulated by a strong enough gravitational wave. Its sensitivity was, however, too low by a factor of about 100,000 for any hope of success. Several gravity wave detectors are currently under construction that should have sufficient sensitivity for a first detection. They are based upon a Michelson interferometer (see *Fourier transform spectroscope*), and have arms several kilometers long, with multiple reflections to multiply that length by up to a factor of 100. Four instruments may be expected to start working in the next few years: LIGO (USA), VIRGO (France–Italy), TAMA 300 (Japan) and the slightly different GEO 600 (Germany–UK).

grazing occultation

An *occultation* when the more distant object runs just along the edge of the nearer object. In the case of an occultation by the Moon, mountains and valleys on the Moon's edge may cause the occulted object to disappear and reappear several times (cf *Baily's beads*).

grazing-incidence telescope

A telescope used to image objects in the *EUV* and soft *X-ray* regions. *Photons* in those regions would be absorbed by, or pass through, a mirror were they fall perpendicularly onto it. They will, however, be reflected if they are incident onto the surface of a mirror at a very shallow angle. The optical surfaces of a grazing incidence telescope are formed from conventional *conic sections*, but are annular segments far from the focus. They thus resemble the shape of a bottomless barrel that has been sawn in half. The Wolter type I design has a concave parabolic primary mirror and a concave hyperbolic secondary mirror. It is the main design to have been used in practice, having been flown on Chandra, XMM Newton, ROSAT and other spacecraft. The basic Wolter type I design intercepts only a tiny fraction of the incoming radiation, since its *effective aperture* is just a thin annulus. Several such telescopes, feeding a common focus, are therefore usually nested within each other in order to improve the sensitivity.

great circle

A circle on the surface of a sphere that exactly divides the sphere in half. So called because it is the largest (greatest) circle that may be drawn on the sphere's surface. For example, the *celestial equator* and the *ecliptic* are great circles on the *celestial sphere*.

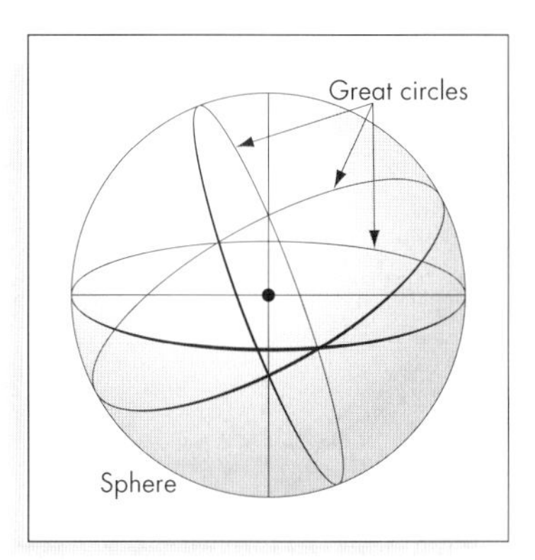

Great circles.

greatest brilliance

The moment when Mercury, or more especially, Venus, is at its brightest in the sky. It occurs at around *elongation* 10° to 12° for Mercury when its magnitude is about -1.2^{m}, and around elongation 35° to 36° for Venus, when its magnitude is -4.6^{m}, and it is the brightest object in the sky after the Sun and Moon. See also: *evening or morning star.*

greatest elongation

The positions of Mercury or Venus when they are at their greatest angular distances from the Sun in the sky. The line of sight from the Earth to the planet is then a tangent to the planet's orbit. Since the orbits are elliptical, the greatest elongation can vary between 18° and 28° for Mercury and between 45° and 47° for Venus. It occurs between 16 and 28 days before/after *inferior conjunction* for Mercury and 70 to 71 days before/after inferior conjunction for Venus. The optimum time for observing these two planets is around greatest elongation, since when they are closest to us (inferior conjunction), we can see only their dark sides. See also: *aspect; opposition.*

Green Bank radio telescope

The largest fully steerable dish-type *radio telescope* in the world, with an aperture 110 × 100 m. Located in West Virginia.

The 110m × 100m Green Bank Radio Dish. (Image reproduced by courtesy of National Radio Astronomy Observatory and Associated Universities Incorporated.)

green flash

A greenish appearance to the top edge of the setting or rising Sun. It is only seen under exceptional observing conditions.

Greenwich mean sidereal time

The local *sidereal time* on the *Greenwich meridian*. It is tabulated for midnight throughout the year in the *astronomical almanac*, and from this the local sidereal time at any other place or time can be calculated.

Greenwich mean time

See *universal time.*

Greenwich meridian

The zero line from which longitude on the Earth is measured. It is the meridian from the north to the south pole that originally passed through the center of the *transit circle* at the *Royal Greenwich Observatory*, London. This telescope was used to make many of the early position measurements upon which early navigation tables were based, and so formed the reference point for those tables. Although the instrument is no longer in its original position, a metal plate set in the ground at the observatory shows where it used to be. The Greenwich meridian was adopted internationally as the reference point for longitude in 1884.

Gregorian calendar

The calendar that is now in very widespread use. It was devised by Christopher Clavius (1537–1612) in order to replace the by then increasingly inaccurate *Julian calendar.* Pope Gregory XIII authorized its use in 1582, by which time the *vernal equinox* was occurring on 11 March, instead of 21 March. Ten days were therefore omitted in order to re-align the calendar and the seasons. The Gregorian calendar, however, was not adopted in the UK and USA until 1752, by when it was 11 days out of step.

The Gregorian calendar has a leap year whenever the year number is divisible by 4, except for century years, which must also be divisible by 400 in order to be leap years. Thus 2000 is a leap year, but 1900 and 2100 are not. The average length of year under the Gregorian calendar is 365.2425 days. The true year length is 365.2422 days. The calendar and seasons hence do continue to get out of step with each other, but it will be the year 4900 AD before the difference amounts to 1 day.

Gregorian telescope

The first design of *reflecting telescope* to be invented, although not the first to be built (see *Newtonian telescope*). The Gregorian telescope was devised by James Gregory (1638–75) in 1663, but it only became popular a century later. The design is similar to the *Cassegrain telescope*, but with a concave elliptical secondary mirror positioned after the focus of the primary mirror. The telescope produces erect images, but its tube length is longer than that of an equivalent Cassegrain telescope, so that it is now rarely encountered.

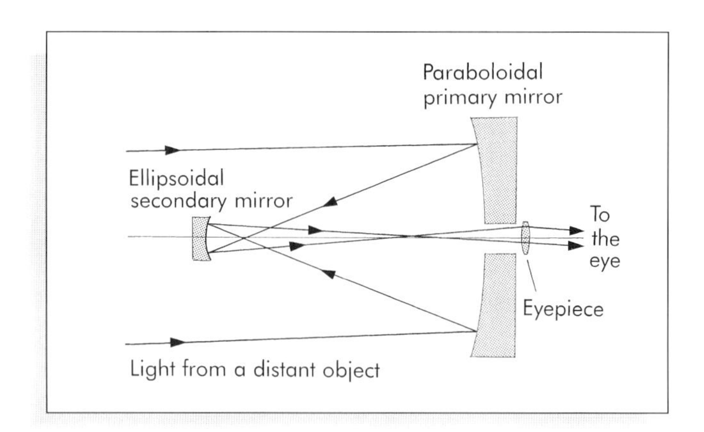

Light paths in a Gregorian telescope.

G

grey scaling

Grey scaling scales the mapping of the intensities in a CCD image for the items of interest to the available display levels on the monitor. For example on a direct solar image, a sunspot might cover intensity levels from 4,000 to 11,000, while faculae cover the range 19,500 to 21,000. Grey scaling to show the sunspots to best effect would then mean setting intensity level 4,000 and below to black (monitor level 0), and intensity level 11,000 to white (monitor level 255), and with the levels covering the sunspots; 4,001 to 10,999, evenly distributed amongst the monitor levels 1 to 254. To show the faculae, would mean distributing the monitor levels 1 to 254 from 19,500 to 21,000, with everything below that range black, and above it white.

Note that in some image processing packages grey scaling may be used for the conversion of a color image to a monochrome image. Other terms for the process may also be encountered such as contrast stretching, level or threshold adjustment. See also: *image processing*.

grinding

A process during the manufacture of optical components when the rough shape required by the component is imparted. *Lenses* are often produced on a lathe, but for small *mirrors*, the process requires two blocks of glass. The first, the *tool*, is fixed firmly to a table or other support. The second, the *blank*, that will become the mirror, is placed on top of the tool with coarse wet grinding powder in between. The grinding powder is usually carborundum since it is harder than glass. The blank is moved back and forth over the tool and becomes concave, while the tool becomes convex. When the approximate shape required for the mirror is reached, the coarse grinding powder is replaced with a finer grade. The process then continues until the pits left by the coarse powder have been removed. A finer still grade of grinding powder then removes the pits left by the previous grade of powder and so on, until the last stage of grinding powder leaves pits that are removable by *polishing*. For large mirrors the tool is usually smaller than the mirror blank, and the latter may also be pre-formed by *spin casting* or by diamond milling. A numerically controled diamond-milling machine may produce a mirror surface ready for polishing directly without the need for some or all of the grinding stages. See also: *figuring*.

grism

A combination of *diffraction grating* and *prism*, sometimes used as the primary disperser of a *spectroscope*, but more often as a *cross-disperser* or in a *direct-vision spectroscope*.

Grotrian diagram

A plot of the *energy levels* of an atom, ion, or molecule.

ground state

The lowest *energy level* of an atom, ion or molecule.

GSC II

See *guide star catalog.*

guide star

A star used for *offset guiding.*

guide star catalog

A catalog of some 15 million stars and 4 million other objects down to a limiting magnitude of $+19^m$. It was produced for use with the *Hubble space telescope.* It may be accessed at http://archive.eso.org/gsc/gsc/. A new version of the catalog, GSC II, is under preparation and expected to be completed in 2001, and to contain details of half a billion objects.

guide telescope

A telescope attached to and aligned with the main telescope. It is usually somewhat smaller than the main instrument, although it will often have a focal length comparable with that of the main telescope. It normally has an illuminated *cross wire eyepiece*, and is used for *guiding* in order to obtain sharp images from long exposures. Some guide telescopes have mountings that enable them to be moved from their alignment with the main instrument so that *offset guiding* may be used on faint or diffuse objects. See also: *finder telescope.*

guiding

During long exposures, the observer uses the *slow motion* adjustments on the telescope's *drives* to keep the object being imaged accurately centered on the cross wires of the *guide telescope.* Errors in the telescope's drive rate, mis-alignment of the axes, atmospheric refraction, movement of the object, etc. are thus corrected, and a sharp image obtained. See also: *finding; Offset guiding; tracking; autoguider.*

gyro-synchrotron radiation

See *synchrotron radiation.*

HA

See *hour angle.*

Hadamard mask imaging

A means of obtaining an image of an extended object using a point *detector.* A series of grids (masks) with differing opaque and transparent sectors are placed over the detector, and the successive outputs recorded. The image is then obtained using the inverse of the matrix representing the masks. In a one-dimensional case (such as spectroscopy), the process is represented mathematically by $I = D\,M^{-1}$, where I is the image vector, D the detector output vector, and M the mask matrix. The technique is most efficient if successive masks are obtained by moving a larger mask over the detector through one row or column at a time. Masks of this type may be generated using Hadamard matrices, hence the name of the technique. See also: *coded array mask.*

Haig mounting

See *barn door mounting.*

halation

A diffuse halo surrounding a bright object that arises from scattered light. The term is especially used in connection with light scattered inside a *photographic emulsion* and which causes photographic images to be larger than they should be.

Hale telescope

A 5m Cassegrain optical telescope with a monolithic honey comb primary mirror, on Mount Palomar, California, altitude 1700 m. See homepage at http://www.astro.caltech.edu/palomarpublic/.

The 5-metre (200-inch) Hale telescope.

Half Moon

The *phase* of the Moon when its visible disk is seen half illuminated. Also confusingly known as first quarter and last quarter, since the Moon is then a quarter or three quarters of the way through its new Moon-to-new Moon cycle. See also: *Full Moon; Old Moon; gibbous phase; crescent phase.*

half-power point

One of two points in the beam from a radio transmitter where the broadcast power has fallen to half the maximum value. For a *radio receiver*, they are the points where the received power from a point source halves as the instrument scans the source. The output voltage from the receiver falls by a factor of the square root of two (to 71% of the peak value) at the half-power points since the power is proportional to the voltage squared.

half-wave dipole

An *antenna* formed from two conducting elements, each a quarter of the operating wavelength long. The elements are in line with each other, and separated by a small distance. The radio signal induces an electric current into the half-wave dipole that is then amplified and converted to the desired form by the remainder of the *radio receiver*. The antenna is most sensitive in a plane orthogonal to the line of the elements, where its *gain* is about 1.6. The sensitivity is zero along the line of the elements. See also: *Yagi antenna.*

H

half-wave plate

A device used in *polarizers* to rotate the direction of *linearly polarized* light. It is made from a birefringent material and designed so that the ordinary and extraordinary rays have a *phase* shift of 180° (half a wavelength) when they emerge. See also: *Pockel's cell; quarter wave plate.*

half width at half maximum

See *full width at half maximum.*

H-α cut-off filter

A *filter* that cuts out wavelengths shorter than about 650 nm, thus allowing through the H-α line at 656.28 nm, but eliminating most *light pollution.* It may be used for imaging gaseous nebulae. Note that it is not the same as the *H-α narrow band filter* and must not be used for solar observing. See also: *light pollution rejection filter;*

H-α filter

A very *narrow band filter* centered on a wavelength of 656.28 nm, and with a *band width* of about 0.05 nm, which isolates the H-α line and so allows direct viewing of the solar chromosphere, flares and prominences. It is based upon the *Fabry-Perot etalon.* (CAUTION – always take appropriate precautions when observing the Sun or damage to the eye and/or telescope can result). See also: *solar filter.*

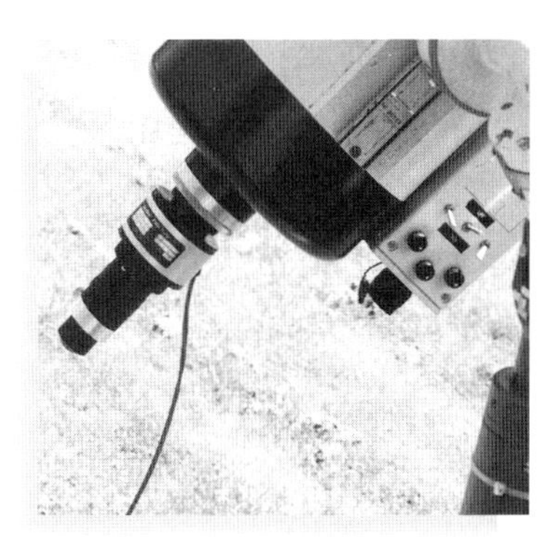

An H-α filter on a small telescope.

H-α prominence telescope

A device similar to a *coronagraph* that also uses an *H-α filter.* It enables detailed images to be obtained of prominences but does not need a high altitude site or the extreme care needed for the successful use of a coronagraph. See also: *prominence spectroscope.*

Handbook of the British Astronomical Association

An almanac and ephemeris published annually by the *British Astronomical Association* containing summaries of planetary positions and astronomical events, etc.

H and K lines
See *Fraunhofer lines.*

hard X-ray region
The short wavelength (high energy) end of the *X-ray region*, extending approximately over the wavelengths 0.4 nm to 0.01 nm.

harmonic frequencies
The frequencies of an oscillating system above the *fundamental frequency*. Harmonic frequencies are usually a simple multiple of the fundamental frequency. See also: *Fourier series.*

Hartebeesthoek radio observatory
A radio observatory with a 26m dish located about 50 km northwest of Johannesburg.

Hartmann formula
An approximate formula for the *refractive index* of an optical material at wavelength, λ, of the form:

$$\mu_\lambda \approx A + \frac{B}{(\lambda - C)},$$

where A, B and C are constants for the material. See also: *Cauchy formula; constringence.*

Hartmann mask
A mask placed over the telescope objective with two or more small holes in it. When *focusing*, especially onto *CCD* chips, the image will show two or more spots so long as it is out-of-focus. When the spots merge, the image is in focus and the mask can be removed to make the main observation. If three holes are used one should be made a different shape from the others, the position of its spot will then indicate whether you are inside or outside the focal point (trial and error is needed to determine which spot position corresponds to being inside the focus and which to outside). See also: *Hartmann test.*

Hartmann sensor
A detector for the deviations arising from atmospheric turbulence in the *wave front* of light from a guide star, used within an *adaptive optics* system. It consists of a grid of small lenses, each of which produces an image of the guide star onto a *CCD*. The movements of those images from their correct positions then provides the details of the wave front distortions. See also: *isoplanatic patch; shearing interferometer.*

Hartmann test
An optical test for a lens or mirror that measures the position of the *focal point* to a high level of precision, and which for mirrors also determines the shape (*figure*) of the surface. A mask with two holes symmetrically placed about the center is positioned over the lens or mirror. Two images of a star are then obtained, one with the detector just inside and one with it just outside the focus. Each will show two images of the star, with their separation proportional to the distance of the detector from the true focus. The process is repeated several times, using different separations for the holes in the mask, in order to analyze the figure of the mirror. See also: *Foucault test; Ronchi test; Hartmann mask.*

A mask for the Hartmann test.

Harvard revised catalog
A catalog of some 9000 stars brighter than 6.5^m published in 1908 by the Harvard college observatory. It was later developed into the *bright star catalog*.

Harvard system
See *spectral type*. See also: *luminosity class*.

Haute Provence observatory
An optical observatory at Forcalquier in the French Alps, altitude 700 m.

haversine
A subsidiary trigonometrical quantity, used in the past before calculators were available, but now rarely encountered. It is defined as:

$$hav\theta = \frac{1}{2}(1 - \cos\theta) = \sin^2\left(\frac{\theta}{2}\right).$$

Haystack observatory
A radio observatory with a 37m dish at Westford, Massachusetts.

H-β filter
See *nebula filter*.

HD catalog
See *Henry Draper catalog*.

heliacal rising
The *rising* of an object just before sunrise.

heliacal setting
The *setting* of an object just after sunset.

helical focuser
See *focuser*.

heliocentric coordinates
Coordinates (e.g. *right ascension* and *declination*) as they would be seen from the center of the Sun. See also: *geocentric coordinates; topocentric coordinates*.

heliocentric Julian date
See *Julian date*.

heliocentric time
The time of an event as seen from the Sun. The terrestrial time has to be corrected for the flight time of the light from the Earth to the Sun (equation of light). It is mostly used to give consistent timings for variable stars, etc. by removing the effects of the changing position of the Earth in its orbit. The correction from the terrestrial time, in seconds, is given by:

$$500(\sin\delta \sin\delta_{Sun} - \cos\delta \cos\delta_{Sun} \cos(\alpha_{Sun} - \alpha)),$$

where α and δ are the *RA* and *Dec* of the object being observed, α_{Sun} and δ_{Sun} are the RA and Dec of the Sun at the same moment.

heliographic coordinates

Latitude and longitude on the Sun – see *Carrington rotation* for the definition of 0° of solar longitude.

heliometer

An obsolete device for measuring the diameters of the Sun and other extended objects and the separations of double stars. It is a *refracting telescope* with the *objective* lens split in half along a diameter. The two halves can be moved along the line of the split. Two images are seen in the *eyepiece*, and the two halves of the objective are then moved until one edge of the object (or one star in the double) is superimposed on the other. The distance between the objective halves is then a measure of the diameter of the object, or separation of the stars.

H

Helios

A commercial firm manufacturing telescopes and accessories.

helioseismology

The study of the solar interior from observations of oscillations of the solar surface.

heliostat

A sideriostat (see *coelostat*) dedicated to solar observing.

Henry Draper catalog

A catalog of the *spectral types* of stars. Produced at the beginning of the twentieth century, principally by Annie Cannon (1863–1941) of the Harvard college observatory. With its extensions it contains entries for some 350,000 stars.

Herschel catalogs

Several catalogs of non-stellar objects produced between 1786 and 1847 by Sir William and Sir John Herschel, and containing some 6500 objects. The objects are almost all included in the *New General Catalog*, and are generally referred to using their NGC numbers. However, occasional reference may be found to objects using these catalogs, indicated by H (Sir William) and h (Sir John), as in H 1.210 and h 2409.

Herschelian telescope

An obsolete form of *reflecting telescope* employed by William Herschel (1738–1822) and others that has its *primary mirror* tilted so that its *focal point* is to the side of the incoming light beam. It can therefore be used without the need for a *secondary mirror*. It was devised when telescope mirrors were made from *speculum metal* and so reflected only 60% of the incoming radiation at best. Telescope designs, such as the *Newtonian* and *Cassegrain*, would thus deliver only about 36% of the incoming light to the observer, compared with 60% for the Herschelian . But tilting the mirror introduces severe *aberrations*, and so the telescope has poor image quality. There are, however, modern variants of the design that use a spherical mirror and correcting lenses, to produce good off-axis images (see *Hobby–Eberly telescope*).

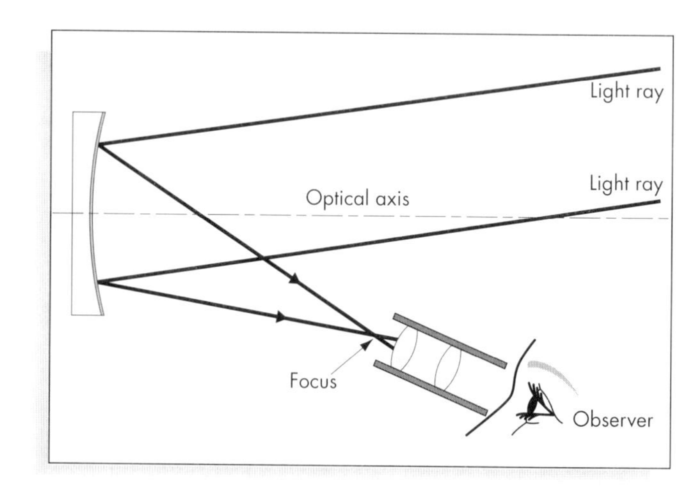

Light paths in a Herschelian telescope.

Herschel wedge

A device that at one time was employed to permit visual observation of the Sun through a small telescope. It is placed immediately before the eyepiece and uses a plain (un-aluminized) diagonal optical *flat* to reflect light through 90°. The second surface of the flat is at an angle to the first (i.e. it is a thin wedge shape) so that the second reflection does not enter the eyepiece. Plain glass reflects about 5% of the incident light, so that the solar intensity is reduced by a factor of 20 or so. Its use is no longer recommended since it does not reduce the solar surface brightness to safe levels unless used on a very small telescope (aperture ≤ 50 mm) and at high magnifications (≥ ×300). Also the un-reflected light beam emerges from the end of the device where it can cause burns on skin and scorch clothes or equipment.
CAUTION – always take appropriate precautions when observing the Sun or damage to the eye and/or telescope can result. See also: *solar filter.*

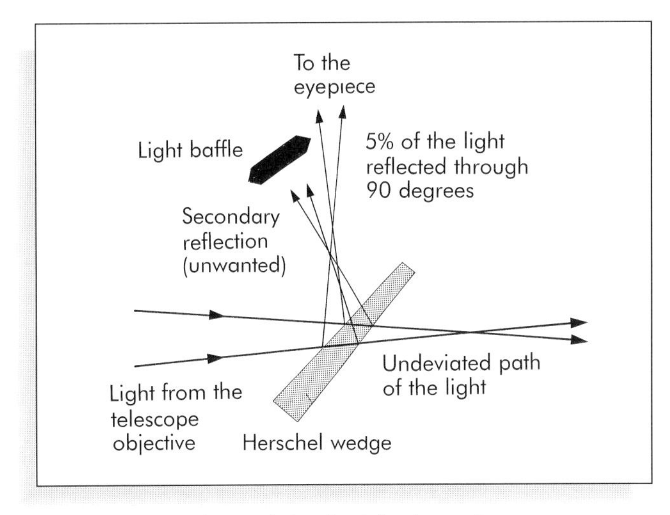

The Herschel wedge (solar diagonal).

hertz

The basic SI unit of frequency, symbol, Hz. It is named for Heinrich Hertz (1857–94). 1 Hz = one cycle of an oscillating system per second.

Hertzsprung-Russell diagram.

A plot of the *spectral type* (or (B–V) *color index* or temperature) against *absolute magnitude* (or luminosity) for a group of stars. Some 90% of the stars lie within a narrow band, known as the main sequence, running from top left to bottom right on the diagram. Most of the remaining stars are white dwarfs in the bottom left hand corner, and there are a small number of giants and supergiants towards the top and right hand side. See also: *two color diagram.*

high pass filter

See *filter.*

Hipparcos catalog

A catalog of the positions, distances, proper motions and magnitudes of some 118,000 stars produced by the *Hipparcos spacecraft*. The uncertainties in its data are about 0.0007″. It may be accessed at http://archive.ast.cam.ac.uk/hipp/. See also: *Tycho catalog.*

Hipparcos spacecraft

A spacecraft, operating from 1989 to 1993, that measured the positions, parallaxes and proper motions of more than 100,000 stars to an accuracy of 0.002″ or better. See also: *Hipparcos catalog; Tycho catalog.*

H line

See *Fraunhofer lines.*

Hobby–Eberly telescope
An optical telescope used for *spectroscopy* at the McDonald observatory, Mount Fowlkes, Texas, altitude 2000 m. The main mirror is spherical and is 11×10 m in size. It is formed from 91 segments kept in place by *active supports*. The telescope points to a fixed *zenith distance* of 35°, and can be rotated fully in *azimuth*. Exposures of up to 2.5 hours can be obtained by moving the detector assembly around the focal plane. See homepage at http://www.as.utexas.edu/mcdonald/het/. See also: *Arecibo radio telescope; SALT.*

The Hobby–Eberly telescope. (Image reproduced by courtesy of Tom Sebring, © Hobby–Eberly telescope.)

H

Homestake neutrino detector
See *neutrino detector.*

homological transform design
A design for a dish-type *radio telescope* in which the shape of the dish is allowed to distort under the changing gravitational load as the telescope moves to observe different parts of the sky. The shape of the dish, however, remains *paraboloidal* at all times, although the individual paraboloids change. The *feed* may need to move to follow the changing *focal position.*

honeycomb collimator
See *collimator.*

honeycomb mirror
A mirror, especially the *primary mirror* of a large telescope, that is thick enough to maintain its shape under gravitational stresses, but whose weight is much reduced by hollowing out the back. The hollows are often in a six-sided honeycomb-type pattern. See also: *segmented mirror; spun cast mirror; thin mirror.*

horizon
1 The actual junction between the Earth and sky, including the effects of altitude and of obstructions such as trees and buildings, known as the true horizon. See also: *dip of the horizon.*
2 The *great circle* on the *celestial sphere* that is perpendicular to the line joining the *zenith* and *nadir*. This will rarely coincide with the true horizon, but it is used to calculate the *rising* and *setting times* for objects in the sky. The predictions of such times will therefore vary slightly from the actual times for a specific observer.

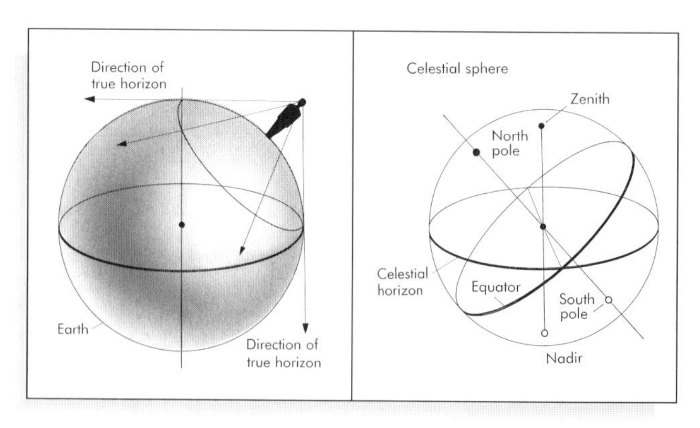

The true horizon.

The astronomical (celestial) horizon.

horizontal coordinates
A synonym for *altitude* and *azimuth.*

horizontal parallax
See *diurnal parallax.*

horn antenna
A horn with a square cross-section that is used to receive short wave radio signals. It may simply be a flared end to a waveguide used as the *feed* at the focus of a paraboloid dish, or it may be sufficiently large that it gathers the radiation directly as with the Holmdel *antenna* that first detected the microwave background radiation.

horseshoe mounting
A design of *equatorial telescope mounting* with the top bearing of the polar axis formed from a large circular annulus. The annulus has a sector missing so that it resembles a horseshoe in shape. The design is very stable and enables the telescope to point to all parts of the sky. It has principally been used for the 5m *Hale telescope.*

H

hour
Apart from being the familiar unit of time, it is also used as a measure of angle. 1 h = 15°. This arises because objects in the sky move through 15° in one hour of *sidereal time*. See also: *minutes; seconds.*

hour angle
The angle, westward from the *prime meridian*, of an object in the sky, measured in *hours, minutes* and *seconds.* See also: *right ascension; sidereal time.*

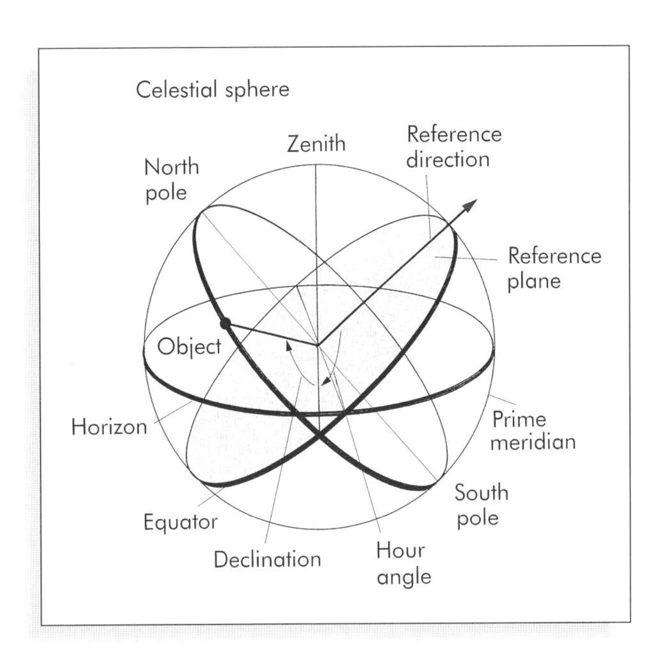

The definition of hour angle.

HPBM
The beam width between the half-power points. See also: *beam width.*

HR diagram
See *Hertzsprung–Russell diagram.*

HR number
See *Harvard revised catalog.*

HST
See *Hubble space telescope.*

Hubble class

A system of classification of galaxies based upon their shapes devised in 1925 by Edwin Hubble. There are three main classes; spiral, elliptical and irregular galaxies, with the spirals sub-divided into normal and barred. The classification is often shown pictorially as the tuning-fork diagram.

Hubble data archive

An archive of some 100,000 images of 20,000 objects observed by the *Hubble space telescope*. It may be accessed at http://www.stsci.edu/archive.html.

Hubble guide star catalog

See *guide star catalog*.

Hubble space telescope

A 2.4m *Ritchey–Chrétien telescope* launched into space in 1990 and operated by *NASA* and *ESA*. It is able to provide high-resolution direct images and spectra over a region from the near infrared to the ultraviolet. See http://oposite.stsci.edu/pubinfo/Pictures.html for images from the telescope.

The Hubble space telescope. (Image courtesy of NASA)

Hutech corporation

A commercial firm manufacturing telescopes and accessories.

Huyghenian eyepiece

A class C (see *eyepieces*) eyepiece made from two *plano-convex lenses* with the *field lens focal length* twice that of the *eye lens*, and a separation half the sum of the two focal lengths. It has a useable *field of view* of about 25°, *chromatic aberration* is small or negligible but *field curvature* is strong. The focal point of the eyepiece is between the lenses, so that it is difficult to add *cross wires*. See also: *Dollond eyepiece*.

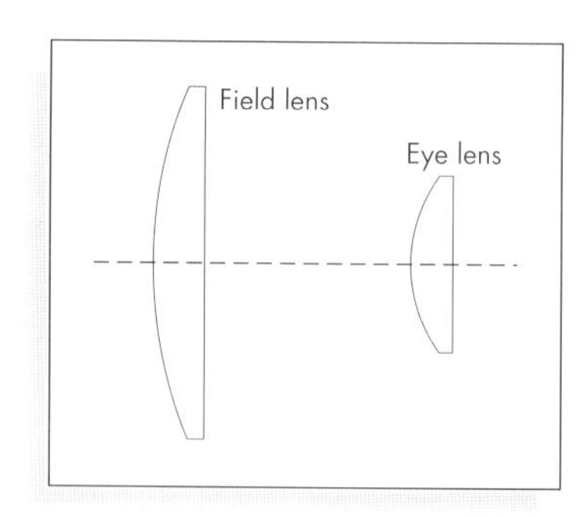

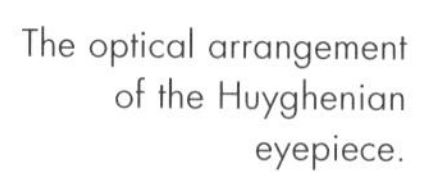

The optical arrangement of the Huyghenian eyepiece.

HWHM

See *full width at half maximum*.

hyperboloid

A shape obtained by rotating a hyperbola around its center line. Mirrors with hyperboloidal shapes are used as the secondary mirrors in *Cassegrain telescopes*, as the primary and secondary mirrors in *Ritchey–Chrétien telescopes*, and in *grazing-incidence telescopes*. See also: *paraboloid; conic section*.

hypering

See *hypersensitization.*

hypersensitization

A collection of processes used to increase the sensitivity of *photographic emulsion*, now of less significance since *CCD* detectors are intrinsically very much more sensitive than photography. The theoretical basis for hypersensitization is poorly understood. Amongst many processes that have been used with success there are; baking at 50 °C to 75 °C for several tens of hours, immersing in nitrogen or *forming gas* (a mixture of nitrogen and hydrogen), exposing to a vacuum and for infrared-sensitive emulsions, bathing in water or dilute ammonia.

Hz

See *hertz.*

I

IAC
See *Instituto de Astrofisica de Canarias.*

IAU
See *International Astronomical Union.*

IBVS
See *Information Bulletin on Variable Stars.*

IC
See *Index catalog.*

IDA
See *International Dark Sky Association.*

illuminance
See *irradiance.*

illuminated cross wire eyepiece
See *cross wire eyepiece.*

image
A pictorial representation of one or more aspects of an object. In astronomy a telescope usually produces the original image. The image can have originated at any wavelength or range of wavelengths. It may then be detected directly with a CCD detector, photographic emulsion, etc. or built up by scanning or through more complex processes such as *aperture synthesis.*

image compression
An algorithm that enables an electronic image to be stored using a reduced amount of memory. There are three main varieties of compression that retain all the information that is within the original image (loss-less compression). They are:

- storing the intensity differences between adjacent pixels rather that the absolute values
- storing runs of similar values as the value times the number of repetitions
- storing runs using a code.

Compressions of factors of 5 or 10 may be achieved by these methods, but will generally be less on noisy images.

Lossy compression, where not all the original information is retained, may be useful with noisy images, since it tends to be the noise that is lost when using these compression algorithms.

image de-rotator
A device to compensate for the rotation of the field of view with time that occurs with *alt-azimuth mounted* telescopes or at the *Coudé focus* of an *equatorially mounted telescope*. See also: *Nasmyth focus.*

image dissector
A device used to split the image of a star into short aligned sections so that all the energy from the star may be fed through a *spectroscope slit.*

image format
See *electronic image format.*

image intensifier
An electronic device the increases the brightness of an image. It consists of an evacuated tube with windows at each end. The front window is coated with a photoemitter (a substance that emits electrons when exposed to light), and the rear window with a phosphorescent material. There is a high voltage applied between the two windows, and there are additional electrodes or magnetic coils to focus the electrons. When the front window is illuminated, photoelectrons are emitted, and accelerated down the tube to collide with the phosphor. The energy gained as the electrons are accelerated means that many photons are emitted by the phosphor for each incoming photon. A single stage image intensifier will produce an image that is brighter by around a factor of ×100. Often three stages are combined giving an amplification of up to ×1,000,000. This amplification, however, is deceptive because noise is amplified as well as the signal. The true gain in efficiency arising from the use of an image intensifier, is thus the *quantum efficiency* of the photoemitter relative to the quantum efficiency of the detector. For visual and *photographic work* this may amount to a factor of ×10 to ×50. But there is no gain for *CCDs* that already have intrinsic quantum efficiencies better than those of the photoemitters. Since image intensifiers are noisy and often distort the image geometrically, their use in astronomy has declined greatly. See also: *electronographic camera; image photon counting system.*

image photon counting system
An imaging system, now superseded by *CCDs*, that combined an *image intensifier*, TV camera and computer to give a sensitivity 10 to 20 times that of photographic emulsion. The noise inherent in image intensifiers was reduced by detecting the individual bursts of output photons originating from each incoming photon (photon counting) and storing the result within the computer.

image plane
The plane at the focus of a telescope wherein lies the accurately focused image. In some instruments, such as the Schmidt camera, the surface of best focus is curved, so that photographic plates, etc. have to be curved to match if the images are to be in focus over the whole plate.

image processing
The manipulation of the electronic version of an image, using a computer, in order to present the image in the desired form. Although there are many individual techniques to image processing such as *grey scaling, sharpening, smoothing, false coloring*, etc., the technique overall divides into *data processing* and image enhancement. The former is the correction of known faults in the image through processes such as *flat fielding*, correction for geometrical distortion, etc. The latter, applied after data reduction, enhances those parts of the image that are of interest. The same image may be enhanced quite differently for different purposes. For example a direct solar image might contain details of sunspots and faculae, but displaying the dark, high contrast sunspots at their best will be very different from showing the bright, low contrast faculae. See also: *unsharp masking; data analysis; data reduction; dodging; filter.*

image reduction and analysis facility
A large widely used software system used for processing data from large ground-based and space based telescopes.

image rotation

A rotation of the *field of view* that occurs with some telescope mounting designs, such as the *alt-azimuth* and *Coudé*, as the telescope *tracks* objects across the sky. See also: *image de-rotator.*

image scale

The relationship on the detector between angle in the sky and linear distance along the detector. It is given by:

$$\frac{200}{D}\ ''/mm,$$

where D is the focal length of the telescope in meters.

image slicer

A synonym for *image dissector.*

I

image stabilizing binoculars

See *binoculars.*

IMB neutrino detector

A *neutrino detector* in Ohio that works by detecting the *Čerenkov radiation* in a large tank of water resulting from high velocity electrons scattered by neutrinos or through inverse-β decay produced by neutrinos, operated by universities at Irvine, Michigan and Brookhaven. See also: *Kamiokande neutrino detector.*

impedance matching

A process required within electrical circuits in order that power is transferred most efficiently. It is essential that *feeds*, cables and amplifiers, etc. in *radio receivers* are matched, so that signal strength is retained.

impersonal astrolabe

See *Danjon astrolabe.*

inclination

See *orbital elements.*

Inconel filter

A full aperture *solar filter* that uses a coating of a variety of stainless steel called Inconel as its absorbing coating.

independent day numbers

A set of eight numbers that enable the position of an object in the sky to be corrected for the effects of *precession*, *nutation* and *aberration*. They are tabulated for each day in the Astronomical *Almanac*. See also: *Besselian day numbers.*

index catalog

The supplements to the *New General Catalog*, containing over 5,000 objects and published in 1895 and 1908. Objects within these supplements are labelled with their IC number (e.g. IC 417).

index error

See *zero error.*

inferior conjunction

A *conjunction* between the Sun and Mercury or Venus, when the planet is closest to the Earth. Although the planet then has its largest angular size, it is poorly placed for observation, since only its dark side is visible from the Earth. See also: *aspect; greatest elongation.*

Information Bulletin on Variable Stars

An irregularly published listing of data on variable stars from the Konkoly observatory, Budapest, sponsored by IAU commissions 27 and 42. See http://www.konkoly.hu/ IBVS/IBVS.html.

infrared

A form of *electromagnetic radiation*, lying between the *microwave* and *visible* regions. The wavelength ranges from 100 μm to 700 nm and the frequency from 3×10^{12} Hz to 8.6×10^{14} Hz.

infrared telescope

A *reflecting telescope* whose primary purpose is to observe at *infrared* wavelengths. The longer wavelength permits the surfaces of the mirrors to be produced to less exacting standards than those on a telescope for use at visual wavelengths, and hence they may be made more cheaply. See also: *flux collector.*

infrared window

One of several regions in the infrared region where the Earth's atmosphere allows some radiation to reach ground level. The main windows are at 1.25, 1.65, 2.22, 3.6, 5, 10 and 21 μm.

ING

See *Isaac Newton group.*

Institut de radio astronomie millimétrique

A French, German and Spanish organization that operates two millimeter-wave instruments, a 30m dish at 2900 m on Pico Veleta in Spain, and a five-element *interferometer* at 2600 m on the Plateau de Bure in the French alps.

Instituto Argentino de Radioastronomía

A radio observatory with two 30m dishes near Buenos Aires, Argentina.

Instituto de Astrofisica de Canarias

A Spanish organization that operates the *Observatorio del Teide* and the *Observatorio del Roque de los Muchachos.* See also: *Gran Telescopio de Canarias.*

Instituto di Radioastronomia

A radio observatory with headquarters at Bologna, Italy, and several out stations. It has two 32m dishes and a 64m dish under construction plus a 640m array.

instrumental profile

See *point spread function.*

integrated magnitude

The *apparent magnitude* of an *extended object* based upon the total energy emitted by the object. See also: *surface brightness.*

integration

A technique for reducing the relative *noise* level in a signal by averaging or adding together many short exposures (or the equivalent). At its simplest it may just require a long enough exposure that the most intense parts of the signal start to *saturate* the detector.

intensity

Colloquially, a term used for any representation of the brightness of a light source. The precise definition is "the amount of energy in the form of electromagnetic radiation passing through unit area into unit solid angle per unit wavelength (or frequency) interval per unit time".

intensity interferometer

An instrument designed to measure the angular diameters of stars. The operating principle was based on interference at visual wavelengths. The main example was built at Narrabri in Australia, and operated from 1965 to 1972 and measured stellar diameters down to 0.0005".

intercombination line

A *forbidden spectrum line*. Intercombination lines, however, are the "least forbidden" of the forbidden lines having probabilities of occurrence that are 1% to 0.01% those of normal spectrum lines, and so can sometimes be found in normal stellar spectra.

interference

1 A phenomenon occurring for any type of wave, but in particular for *electromagnetic radiation*, when two waves interact with each other. For electromagnetic radiation the electric vectors of the two beams combine. If they are in the same direction in space then they add together, giving constructive interference. If the electric vectors are opposed to each other then they cancel each other out, giving destructive interference. Interference occurs whenever two beams of radiation encounter each other, however, unless they are mutually coherent, the effects of interference will change on the same time scale as the frequency of the radiation, and so for most practical purposes be unobservable. When the beams of radiation are coherent (i.e. of the same frequency and with a constant phase relationship), as for example when they result from the splitting of a single original beam, then a stable interference pattern will be observed. This consists of light and dark fringes. The bright fringes (constructive interference) occur at points where the waves are in phase with each other (i.e. the difference in the lengths of the paths taken by the beams in order to reach that point is an integer number of wavelengths long). The dark fringes (destructive interference) occur where the path differences are half a wavelength plus an integer number of wavelengths different from each other. Interference underlies many effects and techniques in astronomy such as the *Airy disk*, the *diffraction grating*, the *Fabry–Perot etalon* and the *interferometer*.

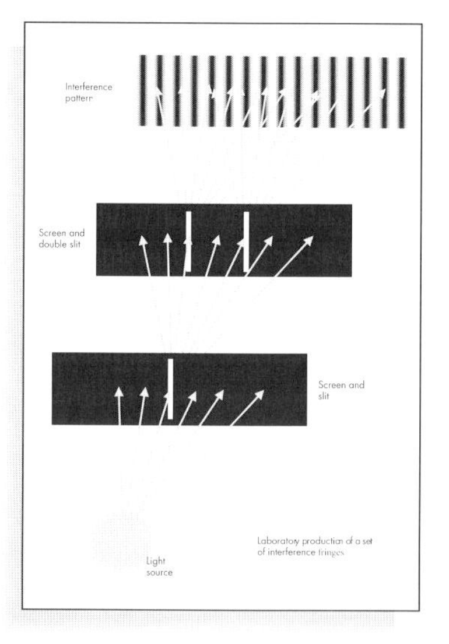

Interference occurring between two light beams originating from a single source.

2 Background *noise*, interfering with an incoming signal, especially at radio frequencies. Often the interfering signal is man made in origin. See also: *light pollution; radio astronomy allocations.*

interference filter

See *Fabry–Perot etalon.*

interferometer

An instrument that uses two or more telescopes to reach high levels of angular resolution. Interferometers have been used at radio wavelengths since the 1950s and they have recently started to be used also for infrared and visual observations (see for example the *Keck telescopes* and ESO's *very large telescope*). The basic operating principle of the interferometer is to combine the outputs from the two telescopes when they are observing the same object. The outputs *interfere* with each other to an extent that depends upon the instantaneous path differences between the signals. As the object moves across the sky, those path differences change, and so the output from the interferometer oscillates. The form of the oscillation may then be used to provide high-resolution data about the object. The angular separation of a double source may be easily determined using an interferometer, but more comprehensive imaging requires the use of *Fourier transforms*. See also: *aperture synthesis; very long base-line interferometry.*

interferometer resolution

The ability of an *interferometer* to separate two point sources. It is given by ($\lambda/2S$) *radians* (where λ is the *wavelength* and S the maximum separation of the aerials or telescopes of the interferometer). See also: *Rayleigh resolution.*

intermediate band photometry

Photometry in the *optical region* that uses filters with bandwidths of 10 to 50 nm. See also: *Strömgren photometry.*

International Astronomical Union

The organization that coordinates much of the astronomical activity around the world. It is based in Paris and founded in 1919. It holds a triennial general assembly whose venue rotates between different astronomically active countries. The general assembly is an occasion for holding many specialist meetings, symposia and colloquiums on a wide range of astronomical topics, but the IAU also sponsors numerous such events on an individual basis at other times and places. The IAU has specialist commissions on all aspects of astronomy, and active researchers will normally want to belong to their relevant commission. The IAU is also responsible for naming of newly discovered objects such as comets and novas, and of the features on planets' and satellites' surfaces. It also operates the *Central Bureau for Astronomical Telegrams* and the *Minor Planet Center.*

international atomic time

Until 1967, the second was defined as 1/31,556,925.9747 of the tropical year for 1900 January 0^d 12^h. Since then its definition is "the duration of 9,192,631,770 periods of the radiation corresponding to the transition between the two hyperfine levels of the fundamental state of the atom caesium 133". Within the limits of measurement, this definition was identical to the one in use previously that was based upon the Earth's orbital motion, and the basis for ephemeris time.

The average results from a large number of atomic clocks around the world are used to provide "International Atomic Time" (Temps Atomique International – TAI), which is now used as the starting point for all other types of time scales. See also: *coordinated universal time.*

International Dark Sky Association

An organization based in Tucson, Arizona, whose purpose is to try to reduce *light pollution*. See http://www.darksky.org/ida/index.html.

interpolation

The use of a graph (or other data set) to predict the value of the quantity between two of the measured or tabulated values. See also: *extrapolation.*

interstellar absorption

The absorption of light by material between the stars, especially dust particles. For lines of sight near the plane of the Milky Way, but not including any dense interstellar clouds, the absorption is by about 0.3^m for every 1000 pc (3,000 ly). In interstellar clouds, however, it can be many orders of magnitude higher than this. The absorption increases rapidly towards shorter wavelengths, causing distant objects to appear redder than they should. This reddening is not the same as the redshift arising from a recessional motion along the line of sight. There is no change of wavelength involved, just a preferential absorption of the shorter wavelengths. See also: *color excess; reddening ratio.*

interstellar extinction

See *interstellar absorption.*

interstellar reddening

See *interstellar absorption.*

Intes-Micro

A commercial firm manufacturing telescopes and accessories.

I

intrinsic color index

See *color excess.*

inverse problem

The problem in non-experimental sciences, like astronomy, of interpreting the observations. Because of *noise* and *uncertainties* in the measurements, and our inability to change the objects observed, there will often be several ways of interpreting data, that may lead to equally good predictions (the usual test of a scientific model or theory). Although there is no definitive solution to eliminating the resulting ambiguity, approaches such as *Ockham's razor* and *maximum entropy* are now widely used.

inverted image

An image that is up side down when compared with the object seen directly. Normal astronomical telescopes produce an inverted image. An additional optical component such as a *relay lens* or *erecting prism* is normally required for *terrestrial telescopes* and *binoculars* in order to provide an *erect image.*

ion

An atom that has a net electric charge as a result of losing or gaining one or more electrons. There are two commonly encountered systems of notation for ions. The first uses "+" or "–" signs as superscripts to the chemical symbol of the atom, with as many of the symbols as there are lost or gained electrons. For example, doubly ionized oxygen would be; O^{2+}, and the negative hydrogen ion; H^-. The second system can only be used for the (commoner) positive ions, and uses a Roman numeral that is larger by one than the number of lost electrons. Thus neutral oxygen is O I, ionized oxygen is O II, doubly ionized oxygen is O III and so on. See also: *ionization; Saha equation.*

ionization

One of several processes whereby an *ion* is produced. The main ones are absorption by the atom or ion of a photon or by its collision with another atom, ion or electron. See also: *Saha equation.*

ionization temperature

A temperature based upon the levels of ionization in a material (see *Saha equation*).

IPCS

See *image photon counting system.*

IRAF

See *image reduction and analysis facility.*

iris diaphragm

A *stop* whose size may be varied. Usually made from a number of flat sheets of thin metal mounted between two concentric rings. Rotation of the rings causes the metal sheets to move towards or away from the center, thus opening or closing the central hole. The iris diaphragm is widely used within *camera* lenses to enable the *depth of focus* and exposure length to be adjusted.

iris photometer

A design of *microdensitometer* that operates by projecting a spot of light of variable diameter onto the stellar image. The diameter of the spot is adjusted until the transmitted intensity reaches a pre-set value, and the diameter of the spot (i.e. the diameter of the iris producing the spot) is a measure of the star's brightness.

I.R. Poyser

A commercial firm manufacturing telescopes and accessories.

irradiance

The total amount of electromagnetic radiation energy falling onto a surface in unit time. Its units are W m^{-2}. The irradiance for just visible light is called the illuminance.

irradiation

A phenomenon of high contrast images whereby the eye sees the bright areas as larger than their real size. It arises through cross-connections between rod or cone cells in the retina, which leads to some of the cells not covered by the bright part of the image nonetheless still sending a signal to the brain. See also: *retina.*

Isaac Newton Group

A part of the *Roque de los Muchachos observatory* that includes the *William Herschel telescope* and the 2.5m Isaac Newton and 1m Jacobus Kapteyn telescopes. See homepage at http://www.ing.iac.es/.

isochromatic emulsion.

A *photographic emulsion* with a more-or-less uniform sensitivity over the *visible region*. See also: *panchromatic emulsion; orthochromatic emulsion.*

ISO number

See *photographic speed.*

isoplanatic patch

The region of sky over which an *adaptive optics* system is able to correct the effects of atmospheric turbulence. It is usually only a few seconds of arc across.

J

James Clerk Maxwell telescope

A 15m millimeter-wave telescope on Mauna Kea, Hawaii, altitude 4100 m.

The James Clerk Maxwell telescope. The protecting membrane that normally covers the instrument has been removed so that the dish itself can be seen. (Image reproduced by courtesy of Robin Phillips, James Clerk Maxwell telescope, Joint Astronomy Center).

jansky

A unit of *flux density*, also called a flux unit, used in radio astronomy, symbol Jy (named after Karl Jansky, 1905–50). 1 Jy = 10^{-26} W m^{-2} Hz^{-1}.

JCMT

See *James Clerk Maxwell telescope.*

JD

See *Julian date.*

Jet Propulsion Laboratory

A *NASA* organization based in Pasadena that tracks and controls spacecraft and runs the *deep space network.*

jiggle photometer

A photographic *photometer* that provides *integrated magnitudes* for extended objects. The photographic plate is moved in a raster pattern during the exposure, so that the images of stars and (say) galaxies are both small squares and so may be directly compared.

jitter

Random variations in a signal. See also: *noise.*

JMI

A commercial firm manufacturing telescopes and accessories.

Jodrell Bank

See *Nuffield Radio Astronomy Laboratory.*

Johnson–Morgan photometric system

See *UBV photometric system.*

Johnson noise

See *noise.*

Johnson Space Center

A *NASA* organization based in Houston that tracks and controls manned spacecraft.

Journal of the British Astronomical Association

A monthly journal published by the *British Astronomical Association* for its members.

Josephson junction detector

See *superconducting tunnel junction detector.*

JPEG

See *electronic image format.*

JPL

See *Jet Propulsion Laboratory.*

Julian calendar

A solar-based calendar devised by Sosigenes (circa 90 BC– ?), and introduced in 46 BC by Julius Caesar. The calendar has a leap year whenever the year number is divisible by 4. The average length of year under the Julian calendar is 365.25 days. The true year length is 365.2422 days. Thus the calendar and seasons get out of step with each other by 1 day every 128 years. It was replaced by the modern system, the *Gregorian calendar*, in 1582 (1852 in UK and USA).

Julian date

A running day number that provides a simple way of calculating the time interval between two calendar dates. The Julian date starts at midday so that there is no change of date throughout the night, and began at midday on 1 January 4713 BC on the *Julian calendar* (24 November 4714 BC on the *Gregorian calendar*). Midday on 1 January 2000 saw the start of JD 2451545. Times other than midday are shown as decimal days. The Julian day number is the whole number part of the Julian date. The heliocentric Julian date is the Julian date corrected for the light travel time difference between the Earth and Sun. See the table of Julian dates in the Appendix (also tabulated annually in the Astronomical *Almanac*). See also: *modified Julian date.*

K
See *kelvin.*

Kamiokande neutrino detector
The University of Tokyo's *neutrino detector* at Kamioka, Japan. Kamiokande II was originally designed to try and detect the decay of protons, and comprised a tank containing 3000 tonnes of water buried underground. It detected neutrinos via the Čerenkov radiation from high velocity electrons scattered by neutrinos or through inverse-β decay produced by neutrinos. Super Kamiokande containing 50,000 tonnes of water has now replaced the first detector and has recently shown that *neutrinos* oscillate between their three types, and so must have a mass. See also: *IMB neutrino detector.*

The interior of the Super Kamiokande neutrino detector during construction. (Image reproduced by courtesy of ICRR (Institute for Cosmic Ray Research), The University of Tokyo.)

Keck telescopes
See panel (page 133).

Kellner eyepiece
A class B (see *eyepieces*) eyepiece made from two *plano-convex lenses.* It is an improved version of the *Ramsden eyepiece* with the *eye lens* formed from an *achromatic doublet.* It is therefore sometimes called an achromatic Ramsden eyepiece. It has a useable *field of view* of 40° to 45°, and the *eye relief* is fairly long. *Chromatic aberration* is almost absent and *field curvature* is reduced compared with the Ramsden eyepiece.

It is the eyepiece design of choice for many observers since it provides a good compromise between performance and price, but it is not suitable for use on telescopes with *focal ratios* shorter than about f6. Some variants on the design using low dispersion glasses for the lenses and known variously as the modified achromat, RK, or RKE designs can be used at slightly shorter focal ratios.

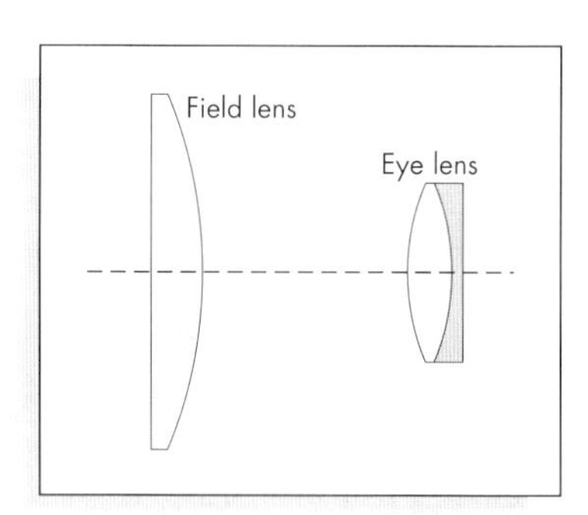

The optical arrangement of the Kellner eyepiece.

kelvin
The SI unit of temperature, symbol K. Named after Lord William Kelvin (1824–1907). The kelvin scale starts at absolute zero and has values of 273. 16 K and 373.16 K at the freezing and boiling points of water respectively. It is thus equivalent to the Celsius (centigrade) temperature scale re-based to absolute zero. Temperatures must normally be on the kelvin scale before they can be entered into physical formulas.

Keck telescopes

Two 10m *Ritchey–Chrétien* optical telescopes at Mauna Kea in Hawaii, altitude 4200 m. The main mirrors are formed from 36 hexagonal segments kept in alignment by *active supports*. See homepage at http://www2.keck.hawaii.edu:3636/realpublic/gen_info/gen_info.html.

The Keck I telescope under construction.

A fish-eye view of the Keck telescope. (Image reproduced by courtesy of the W.M. Keck Observatory.)

The domes of the two Keck 10 m telescopes on Mauna Kea. (Image reproduced by courtesy of the W.M. Keck Observatory.)

One of the mirror segments for the Keck telescopes. (Image reproduced by courtesy of the W.M. Keck Observatory.)

Kennedy space center

A *NASA* organization at Cape Canaveral, Florida, concerned with launching spacecraft.

The space shuttle Columbia during roll-out at the Kennedy Space Center. (Image reproduced by courtesy of NASA.)

kinematic parallax

A value for the parallax (and hence the distance) of an object obtained by comparing its linear velocity through space with its *proper motion*. See also: *moving cluster method*.

kinetic temperature

A temperature based upon the thermal velocities of particles in a material.

K

Kirchhoff's laws

Laws of *spectroscopy* published in 1856 by Gustav Kirchhoff (1824–87). They state that an incandescent solid, liquid or high-density gas emits a *continuous spectrum* while a low-density gas emits or absorbs at particular wavelengths only. Thus stellar spectra have *absorption lines* arising from their low-density outer layers superimposed on a continuum that originates from their deeper layers, while gaseous nebulas have *emission line* spectra.

Kitt Peak observatory

An observatory sited on Kitt Peak in Arizona, altitude 2100 m. It houses the *Mayall telescope*, 3.5m and 2.1m telescopes and numerous other optical and millimeter-wave telescopes operated by several different organizations. See homepage at http://www.noao.edu/kpno/kpno.html.

K line

See *Fraunhofer lines*.

knife-edge test

See *Foucault test*.

König eyepiece

One of a number of class A (see *eyepieces*) eyepiece designs from the same optical designer. They generally have wide *fields of view* (65°) and a performance comparable with the Erfle eyepiece. The most basic design uses just a simple lens and a thick doublet.

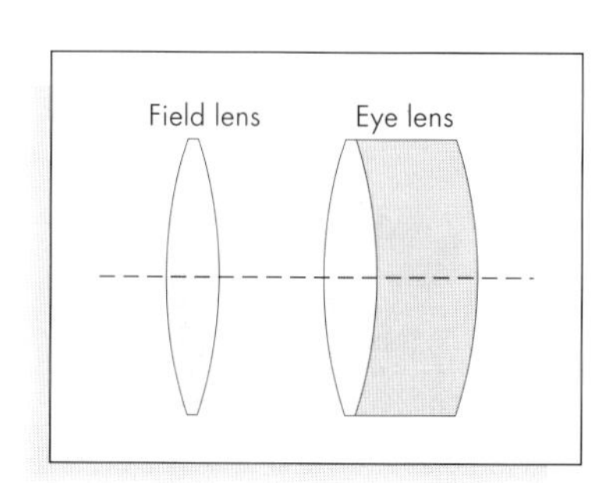

The optical arrangement of the König eyepiece.

Konkoly observatory

An optical observatory in Budapest.

Kottamia observatory

An optical observatory south of Cairo, Egypt, housing a 1.8m reflector.

Krauss radio telescope

A design of *radio telescope* that uses a movable flat reflector to direct the radio waves into a fixed parabolic reflector.

Kueyen

See *very large telescope.*

Kutter telescope

See *Schiefspiegler telescope.*

L

$\lambda/4$ criterion

The standard that the optics of a telescope must achieve, if the instrument is to deliver its *Rayleigh resolution*. The criterion is that the deviations of the *wave front* from perfection must be less than $\lambda/4$. See also: *diffraction limited optics; $\lambda/8$ criterion.*

$\lambda/8$ criterion

The standard that the optics of a reflecting telescope must achieve, if the instrument is to deliver its *Rayleigh resolution*. The criterion is that the deviations of the mirror surfaces from perfection must be less than $\lambda/8$. See also: *diffraction limited optics; $\lambda/4$ criterion; wave front.*

Lanphier shutter

An optical window incorporated into the shutter of a *dome* so that observations can continue with the slot sealed. For use in cold climates so that the interior of the dome may be (slightly) heated, or in hot humid climates to allow cooling and to keep out insects.

lanthanum eyepiece

One of a range of modern class A (see *eyepieces*) eyepieces with five to ten elements and *fields of view* of up to 65°. The *eye relief* is about 20 mm for all focal lengths. The designs are generally variations of the *Plössl* or *Erfle* designs, and are well corrected for aberrations.

lanthanum glass

An optical glass whose composition includes 10% to 20% of the element lanthanum. The glass has a high *refractive index* and a high *constringence* and is used in some eyepiece designs and for some correcting lenses for telescopes.

lap

A layer of pitch on a suitable support that is used in *polishing* and *figuring* a mirror.

La Plata observatory

An optical observatory at La Plata, Argentina. It operates a 2.2m telescope in San Juan.

large binocular telescope

A *Gregorian* optical/infrared telescope on Mount Graham, Arizona. It has two 8.4m monolithic honeycomb mirrors on a single mount, spaced at 14.4 m, which can be used as an *interferometer* to give a *resolution* equivalent to a 22.8m telescope. See homepage at http://medusa.as.arizona.edu/lbtwww/lbt.html.

large zenith telescope

A 6.1m optical telescope using a liquid mercury mirror, currently under construction. It is sited north of Vancouver in British Columbia. The mirror will be spun to produce the required parabolic shape, and it will only point at the zenith, although objects away from the zenith may be observed by moving the detector around the image plane.

Las Campanas observatory
An observatory sited on Cerro las Campanas, Chile, altitude 2300 m. It houses the *Magellan telescopes* and a 2.5m telescope amongst several others.

laser collimator
A low power laser that replaces the *eyepiece* of a telescope during *collimation*. The telescope is collimated when the laser spot is reflected back to its origin.

La Silla observatory
An observatory housing many of the *European Southern Observatory's* instruments, sited on La Silla, Chile, altitude 2400 m.

last quarter
The phase of the waning Moon when its disk is exactly half illuminated. See also: *Half Moon; first quarter; dichotomy.*

latent image
The image within a *photographic emulsion* before it has been *developed*. It is formed from small (10 to 20 atoms) clusters of silver atoms within those silver bromide crystals that have been exposed to light.

latitude
A measure of position on the surface of a spherical object, especially the Earth. It is the angle up or down from the equator to the object. See also: *longitude; astronomical latitude.*

L

LBT
See *large binocular telescope.*

leading edge
The front edge of a moving extended object like the Sun or Moon.

least squares fit
The "best fit" of a linear equation to a set of measurements that contain uncertainties, also known as linear regression. The fit is obtained by minimising the squares of the differences between the data points and the equation. For an equation in the form $y = A\,x + B$, where x and y are the measured quantities, the least squares fit for n data points is given with:

$$A = \frac{\sum_{i=1}^{n}(x_i - \overline{x})(y_i - \overline{y})}{\sum_{i=1}^{n}(x_i - \overline{x})^2} \quad \text{and } B = \overline{y} - A\overline{x}.$$

LEDA
See *Lyon-Meudon extragalactic database.*

lens
A transparent optical component that operates by refracting light. At least one of its surfaces is curved and so light rays hitting it are refracted through differing angles. A beam of parallel light will be brought to a *focus* at a distance from the lens given by the *focal length*. Lenses usually have spherical surfaces, but for some purposes, such as the correcting lens of a *Schmidt camera*, more complex shapes may be needed. A simple lens suffers from *chromatic aberration*, but this may be corrected

by combining two or more lenses made from different glasses to produce an *achromat* or *apochromat*. See also: *converging lens; diverging lens.*

lens speed
See *f-ratio*.

level adjustment
See *grey scaling*.

libration
Any periodic nodding motion of a celestial object, but especially that of the Moon. For the latter, libration arises because the Moon's orbital and rotational periods are identical, but while its rotational angular velocity is constant, its orbital velocity varies because the orbit is elliptical. The rotation thus sometimes "gains" on the orbital motion, and sometimes "loses". As seen from the Earth, the Moon therefore swings East and West over a period of a month by nearly 16°. A smaller movement North and South arises through the Moon's rotational axis not being perpendicular to the plane of its orbit. Through these effects about 59% of the Moon's total surface can be seen from the Earth. See also: *diurnal libration.*

Lick Observatory
An optical observatory belonging to the University of California at an altitude of 1290 m on Mount Hamilton, California. It houses the 36-inch (0.9m) *Lick refractor*, and the 3m Shane reflector. See homepage at http://www.ucolick.org/.

L

Lick refractor
The second largest refracting telescope in the world, with a diameter of 36 inches (0.9m), completed in 1888. It was named for James Lick (1796–1876) who donated the money for its construction. See also: *Yerkes refractor.*

lifetime
The average length of time that an electron will spend in an *excited level* before undergoing a spontaneous *transition* to a lower level. It is the reciprocal of the *transition probability*. The values of lifetimes range from 10^{-8} s for normal transitions to 0.001 s to 10^{7} years or more for *forbidden lines.*

light
See *electro-magnetic radiation.*

light bucket
See *flux collector*.

light curve
A plot of the variation in the brightness of an object with time.

light gathering power
A synonym for *throughput*, See also: *light grasp*.

light grasp
The increased brightness of a *point source* viewed through a telescope when compared with its brightness to the unaided eye. Since the diameter of the *dark adapted* eye is about 7 mm, the light grasp is given by $G \approx 20{,}000D^2$, where D is the *objective* diameter in meters.

light pollution

Unwanted light that is received alongside the light from the object being observed. Light pollution often originates from man-made sources such as street lighting and is one reason why major observatories are sited far from built-up areas. It can also originate naturally from aurorae, scattered moonlight, the zodiacal light, etc. See also: *interference; noise.*

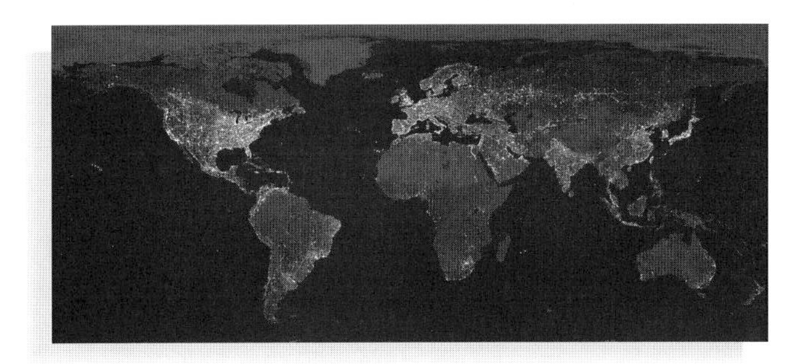

The Earth at night – the artificial lights that lead to light pollution. (Image reproduced by courtesy of Craig Mayhew and Robert Simmon, NASA GSFC, based on DMSP data courtesy Christopher Elvidge, NOAA National Geophysical Data Center.)

light pollution rejection filter

A filter designed to absorb strongly at those wavelengths where the *light pollution* is most intense. If the local street lighting is predominantly low pressure sodium lamps, then a light pollution rejection filter absorbing the region around 590 nm (the sodium D lines) can be very effective. Unfortunately high-pressure sodium and mercury lamps emit light over most of the visible spectrum and so cannot so easily be eliminated. See also: *comet filter; nebula filter; H-α cut-off filter.*

L

light time

A synonym for equation of light (see *heliocentric time*).

light year

The distance moved by light in one year, symbol, ly. 1 ly = 9.47×10^{15} m = 0.306 *parsecs*. See also: *astronomical unit.*

LIGO

See *gravity wave detector.*

limiting magnitude

The faintest *apparent magnitude* that can be seen or detected through a telescope, or the magnitude of the faintest object in a star catalog or on a star map. For a telescope used visually from a good site the limiting magnitude, m_{lim}, is given by; $m_{lim} \approx 16 + 5 \log_{10} D$, where D is the diameter of the telescope's objective in meters. A 0.2m (8-inch) telescope thus has a limiting magnitude of about 12.5^m.

linear regression

See *least squares fit.*

linear response

A response from a *detector* that is directly proportional to the magnitude of the quantity being measured. On a graph, it would be represented as a straight line. See also: *non-linear response.*

linearly polarized radiation

Electromagnetic radiation in which the direction of the electric vector is constant. See also: *elliptically polarized radiation; circularly polarized radiation.*

line broadening
An increase in the width of *spectrum lines*. Line broadening can arise through several processes. The most important is pressure broadening, where the higher the pressure in the gas producing the line, the broader the line becomes. Doppler shifts can also broaden lines leading to thermal broadening from the thermal motions of the emitting or absorbing atoms, and rotational broadening, due to the different velocities of the approaching and receding limbs of the object. Turbulent motions, *Zeeman splitting*, expansion or contraction and many other processes can also broaden spectrum lines. See also: *Doppler broadening*.

line of sight velocity
A synonym for *radial velocity*.

line profile
The precise shape of a *spectrum line*. The line profile is affected by many processes, and potentially can therefore give information on those processes. The effects of the different processes may, however, be difficult to disentangle, and so very high dispersion and resolution spectra are usually needed. See also: *line broadening*.

line receiver
A radio telescope designed to detect spectral lines and therefore with a narrow receiving bandwidth.

line spectrum
A *spectrum* with a *continuum* and superimposed *absorption lines*. Less commonly, an *emission line* spectrum.

LINUX
A computer operating system, available for small computers, that allows the use of very powerful *image processing* packages. See http://www.linux.org/.

liquid mirror
A mirror formed from a reflecting liquid, such as mercury. The liquid is held in a spinning container and its surface takes up a parabolic shape. It can only be used for observing objects close to the zenith, but potentially very large mirrors could be produced in this way. See also: *large zenith telescope; spun cast mirrors*.

Littrow condition
A design for an *achromatic lens* using an equi-convex lens as one of its components. See also: *Fraunhofer lens; Clairault condition*.

local mean solar time
The *hour angle* of the *mean sun* plus 12 hours. It is closely related to *civil time*, but the latter is constant over zones of longitude 15° wide (time zones) for the convenience of every-day living and may also have adjustments for *summer time*, etc. included.

local oscillator
See *radio receiver*.

local sidereal time
See *sidereal time*.

longitude
A measure of position on the surface of a spherical object, especially the Earth. It is the angle between a selected *meridian* (the Greenwich meridian for the Earth) and the meridian through the object. See also: *latitude; Carrington rotation.*

longitude of perihelion
See *orbital elements.*

longitude of the ascending node
See *orbital elements.*

Losmandy
A commercial firm manufacturing telescopes and accessories.

loss-less compression
See *image compression.*

lossy compression
See *image compression.*

Lovell telescope
See panel (page 142).

low pass filter
See *filter.*

Lowell observatory
An optical observatory on two sites at Flagstaff, Arizona, altitude 2200 m. It was founded in 1894 by Percival Lowell (1855–1916)

lower culmination
See *culmination.*

lower transit
A synonym for lower *culmination.*

LPR filter
See *light pollution rejection filter.* See also: *nebula filter.*

Lumicon
A commercial firm manufacturing filters, telescopes and accessories.

luminosity class
A classification of stars on the basis of their luminosity. It depends upon the pressure broadening of spectrum lines. Very luminous stars are many times larger than fainter stars of the same temperature, but the masses of the stars vary by a much smaller amount. The luminous star therefore has low pressure in its outer layers, and so narrow spectrum lines, while the less luminous star has a higher pressure in its outer

Lovell telescope

One of the earliest of the big radio telescopes, operated by the *Nuffield radio astronomy laboratory*. It was originally known as the Mark I telescope. It has a parabolic dish 250 feet (76 m) in diameter and is fully steerable on an *alt-azimuth mounting*. It was completed in 1957, but has had several upgrades and now forms a part of the *MERLIN interferometer*. It is named after Sir Bernard Lovell (1913–) who was the prime mover in getting the telescope built.

The Lovell radio telescope dish. (Image reproduced by courtesy of Ian Morison, Jodrell Bank Observatory and the University of Manchester.)

Inside the Lovell telescope dish. (Image reproduced by courtesy of Ian Morison, Jodrell Bank Observatory and the University of Manchester.)

The Lovell telescope under construction – it was then known as the Mark-1 telescope. (Image reproduced by courtesy of Ian Morison, Jodrell Bank Observatory and the University of Manchester.)

Inside the control room of the Lovell telescope. (Image reproduced by courtesy of Ian Morison, Jodrell Bank Observatory and the University of Manchester.)

layers, and so broader lines. The luminosity class is added as a Roman numeral to the *spectral type*, according to the following scheme:

Luminosity class	Star type
Ia	Bright supergiants
Iab	Supergiants
Ib	Under-luminous supergiants
II	Bright giants
III	Giants
IV	Subgiants
V	Main sequence or dwarfs
VI	Sub dwarfs
VII	White dwarfs.

The Sun thus has a luminosity class V, and with its spectral class, it is a G2 V star. See also: *line broadening*.

lunar eclipse

The darkening of the Moon as it passes into the shadow of the Earth. It is not a true *eclipse*. Lunar eclipses occur at full moon about once every two or three years. Unlike a *solar eclipse*, however, they are visible from the entire dark hemisphere of the Earth, and so appear to be more frequent than solar eclipses. The moon rarely goes completely dark during the total phase of the eclipse, since light is scattered onto it by the Earth's atmosphere (see *Danjon scale*).

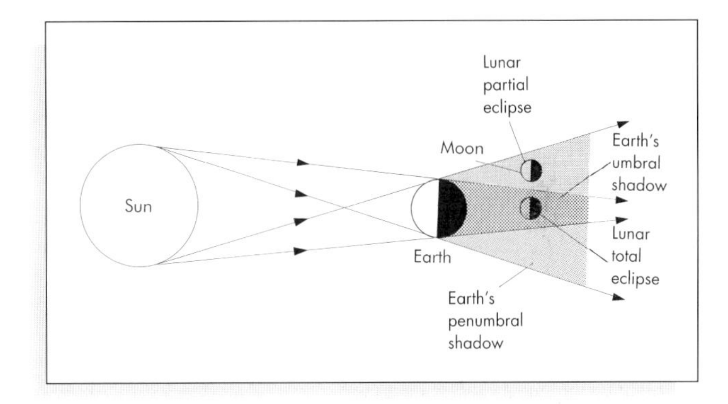

The geometry of a lunar eclipse (not to scale).

lunar parallax

See *diurnal parallax*.

lunation

A synonym for *synodic month*.

luni-solar precession

See *precession*.

LX telescopes

See panel (page 144).

ly

See *light year*.

Lyman ghost

See *ghost*.

Lyman series

See *Balmer series*.

Lyon–Meudon extragalactic database

A database of galaxies, quasars, etc. Available at http://leda.univ-lyon1.fr/. See also: *NASA extragalactic data base*.

LX telescopes

A range of *Schmidt–Cassegrain* and *Maksutov* telescopes produced by the *Meade Instrument Co*. The top of the range LX 200 designs are fully computer controled and have data bases with the positions of some 65,000 objects.

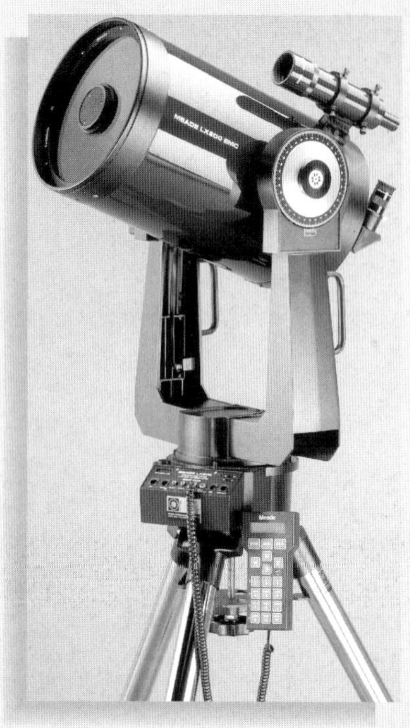

A 10-inch Meade LX 200 Schmidt-Cassegrain telescope. (Image reproduced by courtesy of Meade Instrument Corp.)

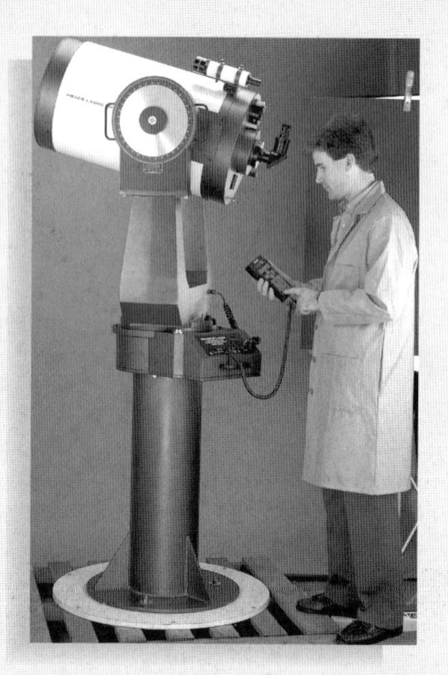

A 16-inch Meade LX 200 Schmidt-Cassegrain telescope on an alt-azimuth pillar mount. (Image reproduced by courtesy of Meade Instrument Corp.)

A 170 mm Meade LX 200 Maksutov telescope. (Image reproduced by courtesy of Meade Instrument Corp.)

Lyot filter

A *narrow band filter* whose operating principle relies on the birefringence of quartz. It is built from a number of units, each of which consists of a plate of quartz and a polarizer. The two in combination lead to constructive *interference* for light passing through at some wavelengths while there is destructive interference at other wavelengths. The wavelengths at which the constructive interference occurs are adjustable by changing the thickness of the quartz plate. Each of the units is therefore designed to allow through the required wavelength whilst cutting out some of the light at unwanted wavelengths. The whole device thus transmits only the required wavelength.

LZT telescope

See *large zenith telescope.*

Magellan telescopes

Two 6.5m optical telescopes for the Mount Wilson Observatory at Las Campanas, Chile. The mirrors are of monolithic *spun-cast* honeycomb construction.

magnetobremsstrahlung radiation

A synonym for *synchrotron radiation.*

magnetogram

An image of the Sun showing the strengths and polarities of the magnetic fields on its surface.

magnetograph

An instrument that measures the intensity and polarity of magnetic fields, especially for the Sun. Its operating principle is based upon the splitting and induced polarization of spectrum lines arising from the *Zeeman effect*. When used in an imaging mode, it produces a *magnetogram*.

magnetometer

M

An instrument for measuring magnetic field intensity and direction directly. For terrestrial use, it may be a simple compass needle, pivoted in the vertical plane, and also able to be rotated in azimuth. A more sensitive magnetometer that is also used on board spacecraft to measure planetary magnetic fields is the flux-gate magnetometer. This is essentially two linked electrical transformers. In the absence of a magnetic field, the currents in the transformers balance each other. But when a magnetic field is present, there is an imbalance in the currents, leading to a net flow, which may be detected and used as to measure the field intensity. Three mutually perpendicular flux gate magnetometers are used when the direction of the field also needs to be known.

Devices that depend upon the Zeeman splitting of spectrum lines may measure magnetic fields on the Sun and stars. The splitting is rarely (except within sunspots) sufficient for the individual Zeeman components to be seen separately, instead the spectrum line is broadened. The Zeeman components are, however, *polarized*, and so they can be isolated by the use of a *polarizer*. On the Sun, fields as weak as 10^{-5} T can be detected, but for stars the limit is about 0.02 T. Interstellar magnetic fields as weak as 10^{-9} T can be detected by the Zeeman splitting of the *twenty-one centimeter line* and other radio spectrum lines.

magnification

1 The increase in the *angular size* of an object when viewed through a telescope. It is given by the ratio of the *objective focal length* to the *eyepiece* focal length. Under normal observing conditions, the useful range of magnifications is from about ×50 to ×500.

2 The increase in the size of the *real image* on a *detector* over that normally given by the telescope brought about by the use of supplementary optics such as in *eyepiece projection*. See also: *minimum magnification; power; chromatic difference of magnification.*

magnified imaging

See *afocal imaging.*

magnitude
See *absolute magnitude* and *apparent magnitude*. See also: *bolometric magnitude*.

Maidanak observatory
An optical observatory on Mount Maidanak in Uzbekistan.

main lobe
The main direction of the sensitivity of a *radio antenna* or *telescope*. Shown on a *polar diagram* as the only or largest lobe. It is the equivalent to the *Airy disk* of an optical telescope. See also: *side lobe*.

major axis
The longest diameter of an ellipse.

Maksutov telescope
A *catadioptric telescope* which uses a spherical *primary mirror*, a *correcting lens* with spherical surfaces, and a *secondary mirror* aluminized onto the inner surface of the correcting lens. The design produces excellent images near the *optical axis*, but *coma* affects the outer part of the *field of view*. The design was invented in 1944 by Dmitri Maksutov, but is actually identical to a design published four years earlier by Albert Bouwers. See also: *ETX telescope*.

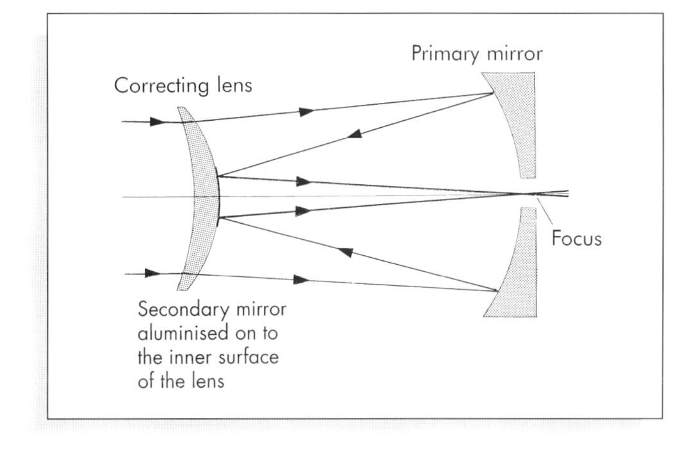

Light paths in a Maksutov telescope.

Mark I radio telescope
See *Lovell telescope*.

maser amplifier
A very low noise amplifier used in some *radio telescope receivers*.

matching
See *impedance matching*.

Maui Space Surveillance System
An optical and infrared observatory operated by the US air force for satellite and asteroid tracking on the summit of Haleakala, Maiu, Hawaii, altitude 3100 m. Its largest instrument is a 3.7m adaptive optics telescope that is sometimes available for astronomical use.

Mauna Kea observatory
An observatory housing numerous large optical, millimeter wave and radio telescopes, including the *Keck telescopes*, *Gemini* north, *Subaru* and the *James Clerk Maxwell telescope*. It is sited on the summit of Mauna Kea, Hawaii, altitude 4200 m. See homepage at http://www.ifa.hawaii.edu/mko/.

maximum entropy method

A mathematical technique that is mostly applied to the interpretation of noisy images. It produces the image that has the least amount of structure within it but which is still consistent with the data. It therefore minimizes the chances of spurious features occurring within the image. The name derives from the concept of an increase in entropy as equivalent to a decrease in the level of structure in a system. The maximum entropy image is therefore the one with the least structure to it. See also: *CLEAN; inverse problem; Ockham's razor.*

Max Planck Institutes

Several institutes in Germany specialising in various areas of science. They have involvement with the *European Southern Observatory* and the radio observatory at Effelsburg in Germany that houses the 100m *Effelsburg radio dish.*

Mayall telescope

A 4m optical telescope at the *Kitt Peak observatory.*

McDonald observatory

An optical observatory sited on two mountains near Fort Davis, Texas, altitude 2000 m. It houses several instruments including the *Hobby–Eberly telescope* and 2.7m and 2.1m telescopes.

McIntosh classification

A three-letter classification scheme for sunspots. The first letter describes the overall appearance of the sunspot or sunspot group, the second describes the largest spot in a group, and the third, the way in which the sunspots are distributed.

M

McMath–Pierce solar telescope

A specialist solar telescope sited at *Kitt Peak.* The telescope is fixed and fed by a *heliostat.*

The McMath–Pierce solar telescope at Kitt Peak.

MDM observatory

An optical observatory on *Kitt Peak.* It houses a 2.4m telescope amongst others.

Meade Instrument Co.

A commercial firm manufacturing telescopes and accessories. See also: *ETX telescopes; LX telescopes.*

mean sidereal time

The *hour angle* of the *First Point of Aries* after the position of the latter has been corrected for *nutation.* See also: *sidereal time; apparent sidereal time; equation of the equinoxes.*

mean solar time
See *local mean solar time.*

mean sun
An imaginary object that moves around the *celestial equator* at a constant velocity and completes a circuit in one year. It was the basis of time keeping until it was replaced by *international atomic time.*

mean value
The average of a set of measurements. See also: *uncertainty.*

Mégantic observatory
An optical observatory on Mont Mégantic, Quebec, at an altitude of 1100 m.

Melipal
See *very large telescope.*

MEM
See *maximum entropy method.*

meniscus lens
A lens with one surface convex and the other concave. See also: *bi-convex lens; bi-concave lens.*

meniscus telescope
See *Maksutov telescope.*

Mercury
An astronomical journal published by the *Astronomical Society of the Pacific.*

meridian
A great circle that passes through the north and south poles on the *celestial sphere.* The meridian that passes through the zenith for an observer is called the *prime meridian* and divides the sky into its eastern and western hemispheres. The *hour angle* of an object is then measured westwards from the prime meridian. Also a similar great circle on the Earth, which represents a line of constant longitude.

meridian circle
See *transit circle.*

MERLIN
A radio *interferometer* operated by the *Nuffield Radio Astronomy Laboratories.* The name derives from Multi-Element Radio-Linked Interferometer Network. It uses seven fixed instruments including the *Lovell telescope* and a 32m dish at the *Mullard Radio Astronomy Observatory.* The longest base line is 217 km, so that its resolution at a wavelength of 10 cm is about 0.05″.

The radio telescope dish at Defford that is part of the MERLIN radio interferometer. (Image reproduced by courtesy of Ian Morison, Jodrell Bank Observatory and the University of Manchester.)

M

Messier catalog
A catalog of 109 fuzzy objects in the sky that the comet hunter Charles Messier (1730–1817) drew up between 1771 and 1781. The purpose of the list was so that he did not waste his time observing the objects in mistake for comets. However, many of the most interesting objects in the sky are now known to be on the list, and so it has become a source of objects that warrant special attention. Objects are designated by M plus the running number, for example the Crab nebula is M1. See Appendix for full list. See also: *Caldwell catalog.*

metal-on-glass mirror
The modern type of astronomical mirror (cf *speculum metal* mirrors). Glass or other material such as *Cer-Vit* or *ULE* is just the support for the reflecting skin. The glass, etc. has its front surface *figured* to the required shape and then *aluminized.*

metastable level
An *energy level* within an atom or ion that has only *forbidden transitions* to lower levels. An electron reaching that level is therefore likely to be trapped there. In some cool molecular clouds the resulting over-population of metastable levels leads to naturally occurring masers.

meteor radiant
See *radiant.*

Metonic cycle
The 19-year interval of time after which the Moon's phases recur on the same dates.

M

Michelson spectroscope
See *Fourier transform spectroscope.*

Michelson stellar interferometer
An *interferometer* used by Albert Michelson (1852–1931) in the 1920s to measure the angular diameters of half a dozen or so stars. The interferometer used two flat mirrors mounted on a 6m long bar. The bar was in turn mounted on the top of the 100-inch (2.5m) Hooker telescope at *Mount Wilson.* Light from a star was fed by the mirrors into the main telescope and produced an interference pattern at the focus. The separation of the mirrors could be changed, and the distance moved by the mirrors between different interference patterns gave the star's angular diameter.

Michigan radio observatory
A radio observatory at Ann Arbor, Michigan. It houses a 26m dish.

microchannel plate
A detector for the ultraviolet and soft X-ray regions. It comprises a thin plate with numerous small (~0.1 mm) holes through it, and with a high voltage (~2000 V) between its two sides. A photon hitting one side of the plate produces an electron. The electron is then accelerated down one of the holes by the voltage. As it travels down the hole it collides with the sides of the hole, and so produces more electrons. A million or more electrons thus emerge from the bottom of the plate for each incoming photon, and can be detected by a variety of methods.

microdensitometer
A device for measuring the density of a photographic image. The machines usually operate by shining a light through the photograph and measuring the transmitted intensity. Usually also the photograph will be mounted on a table so that it can be scanned in one or two dimensions. A one-dimensional scan is used to produce a

tracing of a *spectrum*. Images from *Schmidt cameras* may be scanned in two dimensions by sophisticated computer-controled microdensitometers such as APM and COSMOS. The positions, types, sizes and classes of the objects may then be analyzed automatically.

micrometer
See *bi-filar micrometer.*

micron
An SI unit of length equal to 10^{-6} m, symbol, μm. See also: *nanometer; angstrom.*

microwave
A form of *electromagnetic radiation*, lying between the *radio* and *infrared* regions. The wavelength ranges from 100 μm to 10 mm and the frequency from 3×10^{10} Hz to 3×10^{12} Hz.

millimeter wave
A form of *electromagnetic radiation*, lying between the *radio* and *infrared* regions. The wavelength ranges from 1 mm to 10 mm and the frequency from 3×10^{10} Hz to 3×10^{11} Hz.

Mills cross
An early type of radio *interferometer* comprising numerous simple *antennas* arranged in two perpendicular lines.

minimum magnification
1 The magnification that produces an *exit pupil* with a diameter equal to that of the eye. Lower magnifications than this will still provide images, but will result in an exit pupil larger than the eye's pupil, so that some of the light gathered by the telescope will be wasted. The *dark-adapted* eye has a pupil diameter of about 7 mm leading to the minimum magnification being $M_{min} \approx 140D$, where D is the diameter of the telescope *objective* in meters.
2 The magnification required for the eye to be able to resolve all the detail provided by the telescope. The resolution of the eye is typically 3′ to 5′. Some individuals may be able to improve on this slightly, but for many people it will be worse, and it also tends to deteriorate with increasing age. If the resolution provided by a telescope operating at its *Rayleigh resolution* is thus actually to be seen, the magnification must be at least $M_{min} \approx 1300D$, where D is the diameter of the telescope *objective* in meters.

minor axis
The shortest diameter of an ellipse.

Minor Planet Center
An organization set up by the *International Astronomical Union* and based at the Smithsonian Astrophysical Observatory to which discoveries of asteroids, etc. must be reported in order for you to be accepted by the astronomical community as the discoverer. It may be contacted if you think you have discovered something (but check very carefully first) at http://cfa-www.harvard.edu/cfa/ps/mpc.html or by searching the web for "MPC".

minute
Apart from being the familiar unit of time, it is also used as a measure of angle. 1m = 15′. This arises because objects in the sky move through 15′ in one minute of *sidereal time*. See also: *hour; second.*

minute of arc

See *arc minute.*

mirror

A component of many instruments operating from the radio region to soft X-rays, that functions by reflecting the radiation. Most mirrors have surfaces that are one or other of the *conic sections*. The *primary mirror* of a telescope is usually a paraboloid, while the secondary may be a hyperboloid, flat or occasionally an ellipsoid. At radio wavelengths, the reflector is a conductor, usually a metal, in the optical region it is usually a coating of aluminum on glass or related material as the substrate. At X-ray wavelengths, the mirror is similar to that at optical wavelengths, except that the radiation hits it at a very shallow angle (*grazing incidence*). In all parts of the spectrum, the mirror surface must be correctly shaped to an eighth of the operating wavelength or better if it is to provide images at the *Rayleigh resolution*. See also: *mirror coating.*

mirror blank

A disk of glass or other material such as *CerVit*, or *ULE*, that will have one surface *ground*, *polished* and *figured* to the exact shape required for a *mirror.*

mirror cell

A part of a telescope that holds and supports a mirror. The mirror cell must keep the mirror in its correct position without applying any strain to it. Small mirrors are frequently mounted on a simple three-point support. Large mirrors, however, often have very complex supports, to ensure that the mirror does not distort under its own weight. With modern large thin mirrors the support is require to apply stress to the mirror in a controled manner in order for it to keep to the correct shape. See also: *active support.*

M

mirror coating

One or more thin layers of material on the surface of a mirror. Coatings have two purposes. The first is to provide the actual reflective surface of the mirror. Aluminum is normally used for this, although silver and other metals may sometimes be used instead. The second purpose of a coating is to protect the reflective layer. Aluminum naturally develops a protective layer of aluminum oxide, but an additional thin coat of silicon dioxide will be beneficial. See also: *aluminizing; silvering; Beral.*

MJD

See *modified Julian date.*

MKK system

See *spectral type;* See also: *luminosity class.*

MK system

See *spectral type;* See also: *luminosity class.*

MMT

See *multi-mirror telescope.*

modeling

A technique in widespread use for attempting to understand the details of many types of astronomical object. A theoretical model of the object is produced based upon the laws of physics, and such details of the object as may already be known. Usually the theoretical model will contain simplifying assumptions. The theoretical model is then used to make predictions about the object and the predictions are compared with the observations. If there is a mis-match between the predictions and the observations, then the model is adjusted and new predictions made. The process continues until the predictions are as close to the observations as possible, and the model is then assumed to represent the object's properties accurately. Most modeling now requires computers to cope with the vast number of calculations involved.

modified achromat eyepiece

See *Kellner eyepiece.*

modified English mounting

A design of *equatorial mounting* for a telescope in which a rigid spar is supported on pivots at each end to provide the *polar axis.* The telescope is mounted to one side of the spar with a counter weight on the other side, and pivots orthogonally to the spar to provide the *declination axis.*

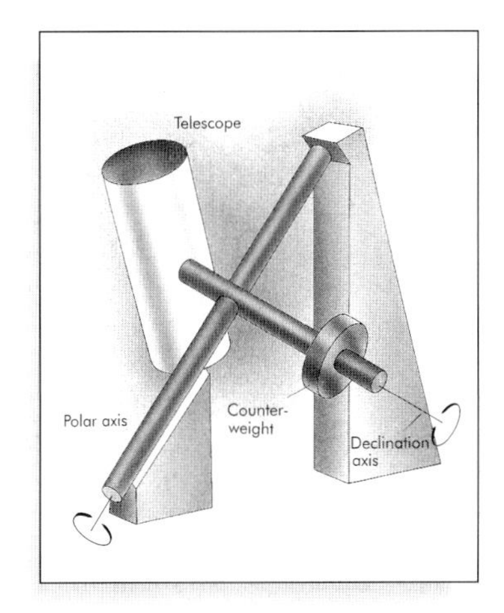

The modified English telescope mounting.

M

modified Julian date

A variation of the *Julian date* that starts at midnight on 17 November 1858. The modified Julian date is thus the Julian date minus 2,400,000.5 days. It is sometimes used for data on spacecraft orbits. (Tabulated annually in the Astronomical *Almanac.*)

modulation collimator

A *collimator* operating at X-ray and gamma ray wavelengths. It comprises two grids made from solid bars that are opaque to the incoming radiation interleaved with clear regions. The grids are separated by a small distance, and have a detector behind them. As the instrument is scanned across the source, the grids alternately obscure and allow through the incoming radiation. The details of the source may then be reconstituted by taking the inverse *Fourier transform* of the output from the detector.

modulation transfer function

A measure of the linear resolution of an optical system or detector, often abbreviated to MTF. Images are obtained of objects that are formed from a grid of sinusoidally varying black and white bars and the objects have a range of spacings for the bars. The modulation transfer is given by:

$$T = \frac{A_r / I_r}{A_o / I_o},$$

where I is the mean intensity, A the amplitude of the intensity variation, the subscript, o, denotes the original object, r, the recorded image. The modulation transfer function is then a plot of T against spatial frequency of the bars in the original objects.

monocentric eyepiece

A class B (see *eyepieces*) eyepiece that is designed to give high contrast by eliminating internal reflections. It is made from three lenses cemented together and with all the surfaces having the same center of curvature. It has a useable *field of view* of only about 25°, and *eye relief* of about 80% of the focal length. It suffers from field curvature and some chromatic aberration. These aberrations are somewhat reduced in the variant of the eyepiece called the triple loupe in which the surfaces' centers of curvature are no longer superimposed. Either design is ideal for planetary or close double star observation.

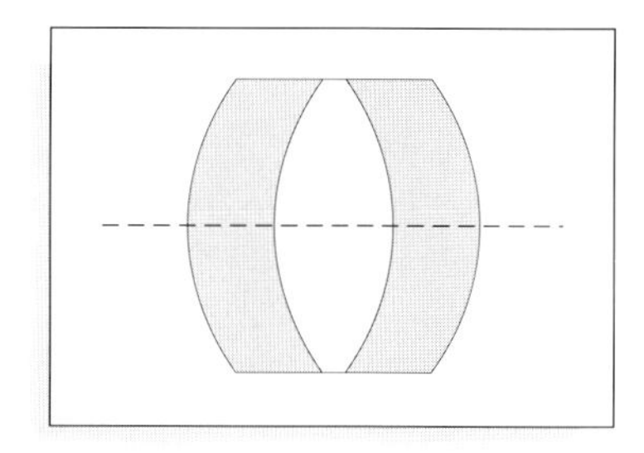

The optical arrangement of the monocentric eyepiece.

monochromatic film

See *black and white film.*

monochromatic vision

The of loss of color in objects at night, that occurs because the rod cells respond only to a single wave band of light (see *retina*).

monochromator

A device for producing light covering a very narrow range of wavelengths. It may also provide an image of the original object at those wavelengths. A simple type of monochromator consists of a low dispersion *spectroscope* with the spectrum imaged onto a slit so that only the selected wavelengths pass through. A variety of *narrow band filters* also act as monochromators. See also: *prominence spectroscope; spectrohelioscope; H-a filter; Lyot filter.*

M

Monthly Notices of the Royal Astronomical Society

A UK-based astronomy research journal published by the *Royal Astronomical Society.*

Mopra observatory

A radio observatory at Coonabarabran, Australia, with a 22m dish.

Morgan–Keenan classification

See *spectral type.*

morning star

Venus around the time of *greatest brilliance*, and when its *elongation* is West of the Sun, so that it is visible in the morning sky before sunrise.

Mount Graham observatory

An optical and millimeter wave observatory, sited on Mount Graham, Arizona, altitude, 3300 m. It is the site of the *large binocular telescope.*

The submillimeter telescope on Mount Graham.

Mount Hopkins observatory

See *Fred Lawrence Whipple observatory.*

mounting

See *alt-azimuth telescope mounting* and *equatorial telescope mounting;* See also: *tripod.*

Mount John University Observatory

An optical observatory near Lake Tekapo on New Zealand's south island.

The Mount John reflector in New Zealand.

Mount Palomar observatory

An optical observatory on Mount Palomar, California, altitude, 1700 m. It houses the 5m *Hale telescope*. See homepage at http://www.astro.caltech.edu/palomarpublic/.

Mount Stromlo and Siding Springs observatory

An optical and solar observatory, near Canberra, Australia, the Siding Springs site houses a 2.3m telescope amongst others and also the *Anglo Australian observatory*.

Siding Springs observatory.

Mount Wilson observatory

An optical, infrared and solar observatory sited on Mount Wilson, California at an altitude of 1700 m. It houses the historically significant 100-inch (2.5m) Hooker telescope built in 1917, amongst several other instruments.

The 100-inch (2.5m) Hooker telescope at Mount Wilson.

M

moving cluster method

A method of obtaining the distance of the stars in a star cluster. The stars in a cluster have a common motion through space. Because of perspective, when seen from the Earth, their proper motions thus appear to diverge from or converge to a point in the sky, called the convergent point. By also measuring Doppler shifts to give the radial velocities of the stars, the distances of the stars may be determined. The method is important since it provides a measure of stellar distances that is independent of parallax, but has so far only been applied successfully to a few star clusters, most notably, the Hyades. See also: *kinematic parallax.*

MPC

See *Minor Planet Center.*

MTF

See *modulation transfer function*

Mullard Radio Astronomy Observatory

A radio observatory near Cambridge, UK. It has played a pivotal role in the development of radio astronomy, seeing the invention of *aperture synthesis*, the discovery of pulsars, and producing several catalogs of radio sources. It also houses the 32m dish of the *MERLIN array*.

multi-coated optics

Optical components, especially lenses and prisms, whose surfaces have received several anti-reflection coatings (see *blooming*) in order to reduce reflection across the whole of the visible spectrum. See also: *mirror coating.*

M

multi-channel photometer

A *photometer* that makes simultaneous measurements through two or more *filters*. See also: *two star photometer.*

multi-mirror telescope

1 A 6.5m optical telescope using a monolithic honeycomb mirror on Mount Hopkins, Arizona. This telescope has replaced an earlier instrument that used six 1.8m mirrors on a single mount, feeding a common focus, from which the name is derived. See homepage at http://cfa-www.harvard.edu/mmt/faq.html.

2 Any telescope, such as the *Keck* and *Hobby-Eberly*, whose mirror is made from several smaller components. See also: *segmented mirror.*

multi-object spectroscopy

A technique for obtaining the *spectra* of many objects simultaneously. *Fiber optic* cables are positioned across the field of view of the telescope, usually by computer-controled arms, to pick up the light from the selected objects, and this is fed to a *spectroscope* with the exit ends of the fiber optic cables aligned along the *spectroscope slit*. The image thus is of a stack of spectra, one for each object.

multiplex advantage

The gain in efficiency of operation for a *Fourier transform spectroscope* when compared with a simple scanning device. The advantage arises because the detector for the Fourier transform spectroscope receives about 50% of the incoming energy on average, compared with 1/n of the energy for the detector of a scanner (where n is the number of measurements made across the spectrum), and thus reaches a given *signal to noise ratio* more quickly. The observing time using a Fourier transform spectroscope is reduced by a factor $\sqrt{\frac{n}{8}}$ compared with the time for a scanner by this effect.

mural quadrant

A device, pre-dating the invention of the telescope, for measuring the *altitude* of a star as it crossed the *prime meridian*. In its typical form, it would be a wall built on a north-south alignment with a quadrant divided into small angles mounted onto it and a rotating arm to point at the star and so to measure its altitude. It was developed to its greatest extent by Tycho Brahe (1546–1601), who was able to measure positions of stars to an accuracy of $\pm$ 30″. See also: *transit circle*.

nadir
The point directly underneath the observer, opposite to the *zenith* in the sky.

Nagler eyepiece
A modern class A (see *eyepieces*) eyepiece made from seven or eight components. It has a useable *field of view* of up to 82°, and the *eye relief* is moderate. *Aberrations* are generally well corrected. The number of lenses involved in the design, however, means that there is significant light absorption, and the eyepieces can be too heavy for some telescopes. Nagler II eyepieces have reduced aberrations.

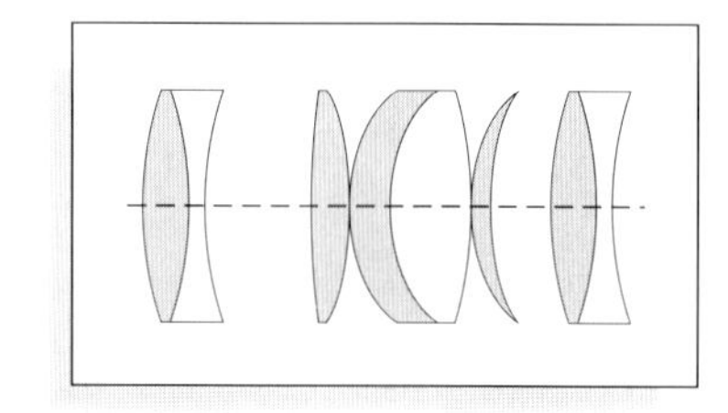

The optical arrangement of the Nagler eyepiece.

Nançay radio observatory
A radio observatory at Nançay, France. It has a 300m long telescope that is a variant of the *Krauss radio telescope* design.

nanometer
An SI unit of length equal to 10^{-9} m, symbol, nm. An alternative unit of length still in widespread use in astronomy is the *angstrom* (10^{-10} m, symbol, Å; 1 nm = 10 Å).

N

Narrabri interferometer
See *intensity interferometer.*

narrow band filter
A *band-pass filter* with a very small *bandwidth.* Typically the bandwidth will be a few nanometers or less. See also: *filter; Fabry-Perot etalon; H-α filter; Lyot filter; monochromator.*

narrow band photometry
Photometry in the *optical region* that uses filters with bandwidths of 10 nm or less.

NASA
See *National Aeronautics and Space Administration.*

NASA extragalactic data set
A catalog of objects outside our own galaxy such as galaxies and quasars. Available at http://nedwww.ipac.caltech.edu/.

NASA infrared telescope
A 3m telescope on *Mauna Kea*, Hawaii.

Nasmyth focus

A focus position for *alt-azimuth mounted* telescopes that is fixed in space irrespective of the telescope's altitude. The system has a plane mirror that reflects the light beam from the telescope down the hollow *altitude* axis. The fixed focus thus emerges from the end of the altitude axis. The focus does move as the *azimuth* of the telescope changes, but on large telescopes a substantial platform may be incorporated on the mounting so that equipment may be built on it and move with the telescope. Most large alt–az telescopes have a Nasmyth focus because it means that bulky, unwieldy, flimsy or delicate equipment can be used since it does not have to tilt with the telescope. The Nasmyth focus has the disadvantages of a small *field of view* that also rotates as the telescope *tracks* across the sky. An *image de-rotator* is therefore needed to provide stationary images. See also: *coelostat; Coudé focus.*

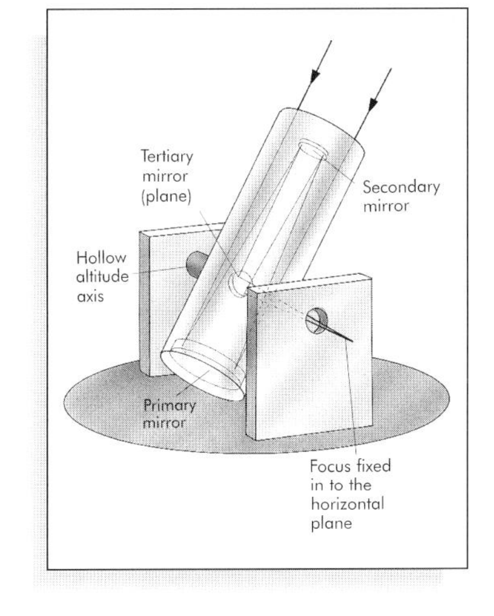

Light paths within the Nasmyth telescope system.

National Aeronautics and Space Administration

A US government organization that oversees civilian aeronautical and space research and development. It is based at Washington, DC, and was founded in 1958. See web site at http://www.nasa.gov/. See also: *Goddard Space Flight Center; Jet propulsion Laboratory; Johnson Space Center; Kennedy Space Center; National Space Science Data Center; Space Telescope Science Institute.*

National Astronomical Observatory

A Japanese organization responsible for several major astronomical facilities including the *Subaru telescope;* the *Nobeyama observatory* and gravity wave detectors.

National Center for Radio Astrophysics

An Indian organization the operates the giant meter-wave radio telescope with thirty 45m dishes and the 530m array at *Ootacamund observatory.*

N

National Optical Astronomy Observatories

An organization overseeing the operation of several large optical observatories and telescopes, including *Cerro Tololo*, *Kitt Peak*, and *Gemini.*

National solar observatory

An organization operating solar telescopes at *Kitt Peak* and Sacramento peak, and managing the *global oscillation network group.*

National Space Science Data Center

A data center for astronomical and space information maintained by NASA at the Goddard Space Flight Center. It may be contacted via; http://nssdc.gsfc.nasa.gov/. See also: *Centre de Données astronomiques de Strasbourg; Astronomical Data Center.*

Nature

A UK-based weekly science and astronomy research journal.

nautical almanac

A version of the astronomical *almanac* produced for navigators.

nautical twilight

See *twilight.*

near field response

The response of an *antenna* for objects close enough to the antenna that the antenna's dimensions must be taken into account. See also: *far field response.*

near infrared

A form of *electromagnetic radiation,* lying at the short wavelength end of the *infrared* region. The wavelength ranges from 700 nm to an ill defined longer wavelength limit variously taken to lie between 5 μm and 40 μm.

near ultraviolet

A form of *electromagnetic radiation,* lying at the long wavelength end of the *ultraviolet* region. The wavelength ranges from about 200 nm to 350 nm.

nebula filter

A *narrow band filter* designed to transmit the strong emission lines from gaseous nebulas. These lines are principally the *forbidden lines* of doubly ionized oxygen at 495.9 nm and 500.7 nm or the H-β line at 485.6 nm resulting in oxygen-III and H-β filters respectively. Use of the filter enables longer exposures to be used, and hence images obtained with improved signal to noise ratios, since much of the *light pollution* is eliminated, but most of the light from the nebula retained. See also: *light pollution rejection filter; comet filter.*

Nebula filters to isolate the oxygen-III lines. (Image reproduced by courtesy of Meade Instrument Corp.)

NED

See *NASA extragalactic data set.*

negative

The original photographic plate or film containing the *negative image* of the object. See also: *photographic emulsion; positive image.*

negative image

An image wherein the brightest parts of the original object are reproduced as the darkest parts of the image, and the darkest parts of the original object are reproduced as the lightest parts of the image. Most often encountered as the original image produced by *photography.* See also: *photographic emulsion; printing.*

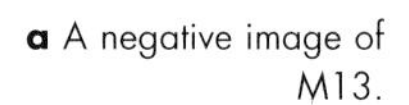

a A negative image of M13.

b A positive version of M13.

negative lens

See *diverging lens.*

NEP

See *noise equivalent power.*

nested telescopes

See *grazing-incidence telescope.*

NESTOR neutrino detector

A *neutrino detector*, based upon detecting the Čerenkov radiation from high-energy electrons produced in water within the Mediterranean.

neutrino

A sub-atomic particle with zero electrical charge, and a mass currently estimated at around 0.0001% that of the electron. Its existence was first suggested in 1930 by Wolfgang Pauli (1900–58) in order to account for the apparent non-conservation of energy in β-decays. It was discovered experimentally in 1955. Its interaction with other forms of matter is very weak. Neutrinos are released during the conversion of hydrogen to helium in stars' centers. They can be observed coming from the center of the Sun, but with only about a third of the expected intensity, giving rise to the solar neutrino problem. The original neutrino is associated with normal electrons and called the electron-neutrino. The other two neutrinos are associated with the heavy electrons (leptons) known as muons and tau particles and so are named muon-neutrinos and tau-neutrinos. Recently evidence has been found to show that the neutrino types can convert from one to another, if true, this would explain the deficit of solar neutrinos.

N

neutrino detector

The first *neutrino* detector started operating in 1968 and continues to this day. It comprises a tank of some 600 tonnes of tetrachloroethene, buried 1.5 km below ground at the Homestake gold mine in Dakota. It is buried deeply in order to shield it from cosmic rays and other forms of radioactivity whose reactions would otherwise overwhelm the neutrino detections. The neutrinos are detected through their conversion of chlorine-37 to argon-37. The latter is radioactive and can be detected as it decays via its emission of a high-energy electron. The very small number of argon atoms produced by neutrinos in the tank are accumulated by purging the tank with helium, and then freezing the argon out of the helium. Two other detectors, *SAGE* and *GALLEX* detect neutrinos using the conversion by a neutrino of gallium-71 to radioactive germanium-71.

More recently neutrinos have been detected via the Čerenkov radiation from high-energy electrons in water. The electrons result from a neutrino and proton converting to a neutron and positron or gain their energy by direct collision between the neutrino and an electron. See also: *Čerenkov detector; IMB neutrino detector; Kamiokande neutrino detector; SNO neutrino detector; DUMAND neutrino detector; NESTOR neutrino detector; AMANDA neutrino detector.*

neutron

A sub-atomic particle that is a fundamental component of the nucleus of an atom. It has a mass of 1.6749×10^{-27} kg, and is electrically neutral. Variations in the number of neutrons in the nucleus give rise to different isotopes of an element. See also: *proton; electron.*

New General Catalog

A catalog of nebulas and star clusters published by Johan Dreyer (1852–1926) in 1888, but based on earlier work by John Herschel (1792–1871). With later published supplements, it lists over 13,000 objects. Many galaxies, nebulas and star clusters are known by their NGC numbers. Thus the Crab nebula is NGC 1952 (and also M1 from the *Messier catalog*). See also: *Index catalog; Herschel catalogs.*

New Moon

The phase of the Moon when it is in *conjunction* with the Sun, and so we see only its dark side. Also the thin *crescent* Moon for a few days after conjunction. See also: *Full Moon; Old Moon; Half Moon; gibbous phase.*

New Moon in the Old Moon's arms

A phenomenon to be observed soon after *new Moon* or just before *old Moon*, when the dark portion of the lunar disk is illuminated sufficiently by reflected Earthlight to be seen faintly cradled by the thin *crescent* Moon.

New Scientist

A UK-based weekly popular science magazine.

new technology telescope

A 3.5m optical reflector at the *European Southern Observatory*. It was the first large instrument to sharpen its images through the use of *adaptive optics,* hence its name.

N

Newtonian–Cassegrain telescope

A *Cassegrain telescope* with an additional flat mirror positioned after the secondary mirror to reflect the light out of the side of the telescope tube. The instrument is usually found as part of a *Nasmyth design*.

Newtonian Telescope

See panel (page 163).

NexStar telescopes

A range of telescopes produced by *Celestron*. The top of the range designs are computer controled and have data bases with the positions of some 18,000 objects.

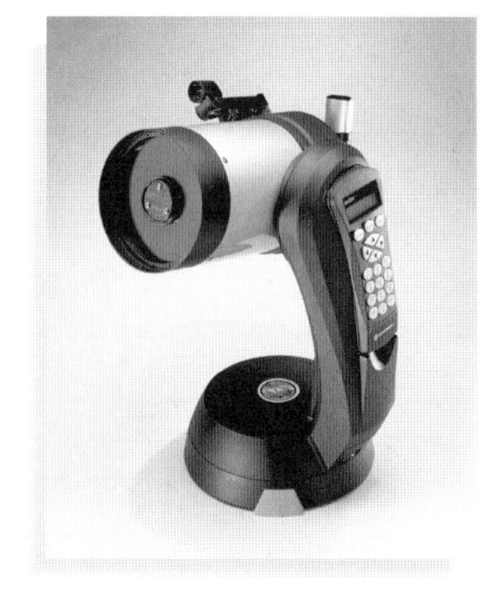

Celestron's NexStar telescope. (Image reproduced by courtesy of Celestron International.)

next generation space telescope

A space telescope planned for launch in 2009. It is expected to have an 8m segmented mirror. See homepage at http://ngst.gsfc.nasa.gov/index.html.

Newtonian Telescope

A *reflecting telescope*, originally designed and built by Sir Isaac Newton in 1668. It uses a *parabolic mirror* as the *objective*. The *secondary mirror* is flat and set at 45° to the optical axis. The light is reflected to a *focus* at the side of the top end of the *telescope tube*.

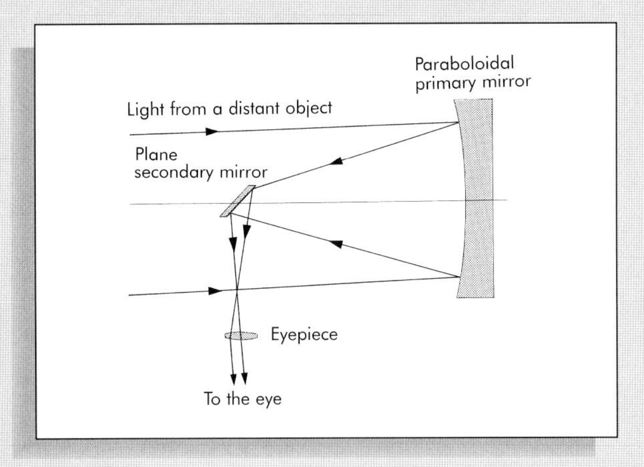

The optical system of a Newtonian telescope.

Newton's original telescope.

Looking down a 13-inch Newtonian telescope with the secondary mirror mount silhouetted against the main mirror.

A modern Newtonian telescope. (Image reproduced by courtesy of Meade Instrument Corp.)

A giant Newtonian telescope – the 'Yard Scope' – with a 36-inch (0.9 m) aperture, and Tom Clark, its owner.

NGC
See *New General Catalog.*

NGST
See *next generation space telescope.*

Nicol prism.
A device for producing linearly polarized light, based upon a calcite crystal. It consists of two right-angled prisms with parallel optical axes, cemented together along their bases. It is similar to the *Glan–Thompson prism*, except that in the latter the prisms are cemented along their long sides. See also: *beam splitter.*

night vision
See *dark adaptation.*

NOAO
See *National Optical Astronomy Observatories.*

Nobeyama observatory
A radio and millimeter wave observatory at Nobeyama in Japan. It has numerous instruments including a 45m dish telescope.

nocturnal
A synonym for a *nocturne.*

nocturne
N

An ancient device for determining the time at night. It was aligned on the pole star and the time read off from a dial by sighting on the positions of stars such as the pointers in Ursa Major.

node
One of two points in a planet or other object's orbit where it crosses the *ecliptic* (or sometimes the *celestial equator*). The node where the object passes in a direction from south to north is the ascending node, the other the descending node.

noise
Any form of spurious signal that interferes with the signal of interest, and so renders the measurement of its value less precise, or even makes it undetectable. There are innumerable sources and types of noise, and no signal or measurement can ever be noise free. The noise sources, however, can be usefully subdivided into four types: intrinsic, extrinsic, signal and processing.

Intrinsic noise originates within the detector, and for example includes thermal noise (also known as Johnson or Nyquist noise) arising from the random thermal motions of electrons and atoms, Shot noise arising from variations in the diffusion rates across the neutral zone of a junction-type detector, chemical *fogging* in *photographic emulsions*, variations in the photon emission efficiency across the cathode of a *photomultiplier*, read-out noise in CCDs, etc.

Extrinsic noise arises from sources and processes not directly related to the instrument or the source being observed. Thus it includes *light pollution*, radio *interference*, thermal emission from the telescope and surroundings at infrared wavelengths, *sky background*, *cosmic ray* spikes on *CCD images*, etc.

Signal noise is noise arising from the nature of the signal. It is principally due to the quantum nature of *electromagnetic radiation*, so that for faint sources, photons arrive in irregular bursts rather than at regular intervals, even though the source may be of constant intensity.

Processing noise arises from the equipment used to process the signal. It can include the noise from amplifiers, etc. but also noise introduced during *data reduction* and analysis. It should be negligible for a good detection system. See also: *fat zero; systematic noise; white noise; random noise; uncertainty; Wiener filter.*

noise equivalent power
A measure of the *sensitivity* of a detector. It is the power of the incoming signal that gives a *signal to noise* ratio of one. See also: *detectivity; normalized detectivity.*

noise reduction
Any process, technique, equipment, etc. used to reduce the effect of *noise* on a signal. There are many approaches to noise reduction, and they vary with the type and source of the noise. Some commonly encountered methods are; *integration*, *cooling*, *chopping*, repeated measurements, *maximum entropy method*, selection of a good observing site and comparison with a stable standard source.

noise temperature
The temperature at which the thermal noise from a simple resistor with the same impedance as an item of electrical equipment equals the noise from that equipment whatever its source.

nomenclature
See *comet nomenclature*, *stellar nomenclature*, *variable star nomenclature.*

non-linear response
An output from a detector that does not vary in direct proportion to the change in the incoming signal. Many detectors have a *linear response* over their useful range, but become non-linear if their performance is pushed to the limit, especially as they approach *saturation.*

normal distribution
See *Gaussian distribution.*

normalized detectivity
A measure of the *sensitivity* of a detector, often symbolized as, D*. It is the *detectivity* multiplied by the square root of the detector area. See also: *noise equivalent power.*

northern lights
See *aurora.*

north polar distance
The angular distance of an object in the sky from the north *celestial pole.* It is equal to 90° – *declination.*

north polar sequence
A set of *standard stars* for measuring *apparent magnitudes* close to the north *celestial pole.*

notch filter
See *filter.*

NSSDC
See *National Space Science Data Center.*

NTT

See *new technology telescope.*

nuclear emulsion

A photographic emulsion with a very high proportion of silver bromide. Used in the form of a thick block to detect cosmic rays via the tracks that their ionization trails leave in the emulsion.

Nuffield Radio Astronomy Laboratory

The University of Manchester's radio astronomy laboratory and observatory, based at Jodrell Bank in Cheshire and founded in 1945 by Sir Bernard Lovell (1913–). Its main instrument is the Mark I 250-foot (75m) fully steerable dish now called the *Lovell telescope*. This telescope along with several others forms a part of the *MERLIN interferometer*.

null test

Any test for which the object under test is correct when the output from the testing apparatus is zero. It is used particularly in testing the surface shape of mirrors. For example the *Foucault test* operates as a null test for spherical mirrors since when the mirror is correctly shaped, the reflected rays are all returned to single focal point, and so the mirror darkens instantly when the knife-edge cuts the returned beam at the focus.

nutation

A small oscillation in space of the Earth's rotational axis arising from gravitational perturbations by the Sun and Moon. The main component of nutation has a period of 18.6 years and a 9″ amplitude, but there are also many smaller components. See also: *precession.*

Nyquist frequency

The highest frequency that will be correctly sampled when a signal is digitized. It is half the sampling frequency. Higher frequencies will not be reproduced correctly but may introduce spurious lower frequency components (see *aliasing*). See also: *sampling theorem.*

Nyquist noise

See *noise.*

object glass

The *objective* of a refractor.

objective

The main light gathering optical component of a *telescope*. In *refractors* it is a lens, in *reflectors*, a mirror, while in *catadioptric telescopes* both lenses and mirrors are used to gather the light.

objective grating

A coarse grating covering the whole *objective* of a telescope. It produces short *spectra* on either side of the main image of a star. Normally the spectra are too short to be useable as such, but they are used in *photometry* to extend the effective *dynamic range* of a photographic plate, since the main image of the star may be *saturated*, while these subsidiary images are not.

objective prism

A large, low angle prism that covers the whole objective of a telescope. It converts each image into a short spectrum allowing many spectra to be obtained with a single exposure.

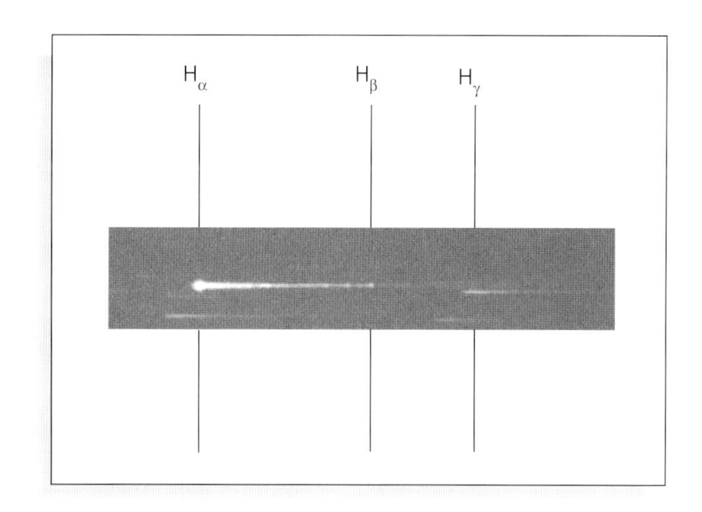

The spectrum of Nova Cyg 1992 obtained using an objective prism. (Image reproduced by courtesy of Bob Forrest, University of Hertfordshire Observatory.)

oblateness

A measure of the degree of non-sphericity of a planet or other object, also known as ellipticity. The non-sphericity or flattening of the planet arises through its rotation. Oblateness is given by:

$$\frac{D_e - D_p}{D_e},$$

where D_e is the equatorial diameter and D_p the polar diameter (sometimes the mean diameter may be used as the denominator). Saturn has the greatest oblateness of any of the major planets with a value of 0.108. See also: *eccentricity*.

obliquity of the ecliptic

The angle between the *ecliptic* and the equator, usually symbolized by ε. It is also the angle of tilt of the Earth's rotation axis. Its value was 23° 26′ 21″ on 1 January 2000.

Observa-Dome

A commercial firm manufacturing telescope domes.

Observatoire de Haute Provence

An optical observatory at St Michel-l'Observatoire, France, altitude 650 m.

Observatorio Astronómico Nacional

A Mexican optical observatory in the Sierra San Pedro Martir, Baja California, it houses a 2.1m telescope amongst several others.

Observatorio del Roque de los Muchachos

An optical observatory on La Palma, Canary islands, at an altitude of 2400 m. It houses many instruments including the *Isaac Newton Group*, the 2.6m Nordic optical telescope and the planned *Gran Telescopio de Canarias*.

Observatorio del Teide

An optical observatory on Tenerife, Canary islands, at an altitude of 2400 m.

observatory

See panel (page 169).

observing chair

A chair especially suited to use while observing. For unaided observations, such as meteor watching, a deck chair is as good as anything. For small telescopes and binoculars, the instrument may be mounted onto a conventional reclining chair, especially if it can be swiveled easily as well. For larger telescopes a chair that is adjustable at least in height by the observer whilst he/she is sitting in it is needed, and this will usually have to be specially designed and made. Telescopes 2 m or more in diameter will usually have an adjustable seat for the observer incorporated into the *telescope mounting*.

occultation

The passage of one object in the sky in front of another, when the nearer object has a much larger angular size than the more distant one. For example, the occultation of a star by the Moon. (Tabulated annually in the Astronomical *Almanac*.) See also: *eclipse; transit*.

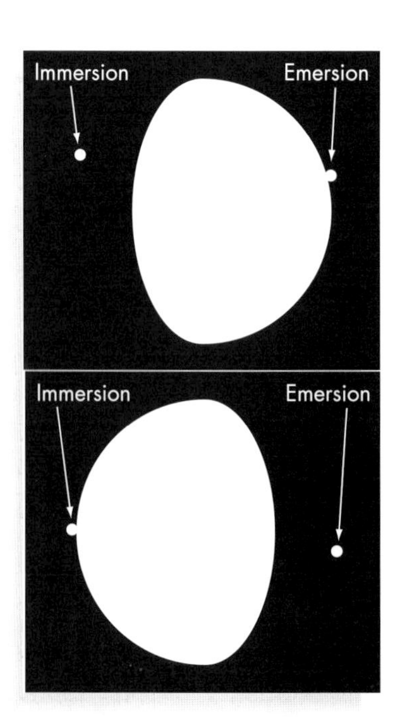

Occultations using waxing (top) and waning (bottom) lunar phases.

observatory

Any site or building used for astronomical observing. Usually the term denotes a building housing a telescope, but it may also be used for a site with many telescopes or for the headquarters of a large astronomical organization, or for a spacecraft used for astronomical observation. Professional observatories now tend to cluster on the few high altitude sites that provide optimum observing conditions (see individual entries for details). Some details of the larger observatories (optical and radio) are tabulated annually in the Astronomical *Almanac*. See also: *dome*.

Alan Heath's back garden observatory.

Part of the Mauna Kea observatory with the Gemini North telescope dome in the foreground. (Image reproduced by courtesy of the Gemini observatory.)

A simple run-off shed to protect a telescope when not in use.

'Domes' do not have to be round!

O

occulting disk
A disk or other solid object placed at the focus of an *eyepiece* and used to obscure (occult) a bright object so that fainter objects may be seen. See also: *coronagraph.*

Ockham's razor
The principle in science that if two theories or models fit the facts equally well, then the simpler one or the one that makes the fewer assumptions should be chosen. See also: *inverse problem.*

octant
A sector that is an eighth of a circle. Also the name of an ancient device used to measure angles between objects in the sky, that consisted of such a sector graduated into small angles and with two sighting rods to align on the stars. The device later developed into the modern *sextant.*

ocular
A synonym for *eyepiece.*

off-axis guider
A device that is placed between the telescope and the *camera*. It projects a small mirror or right-angle prism into the light beam. This reflects a small part of the field of view through 90° into a *cross wire eyepiece*. The telescope is set so that the object of interest is centered on the camera. The off-axis guider is then moved around the field of view until a suitable star is found that may be used for *guiding*. See also: *off-axis guiding; off-set guiding.*

O

off-axis guiding
A technique of *guiding*, used when the main object being imaged is very faint, or diffuse (e.g. a galaxy or interstellar nebula), or your main telescope does not possess a suitable separate *guide telescope*. It uses an *off-axis guider*, and guiding is done using a suitable star close to the object of interest. See also: *off-set guiding.*

An off-axis guider. (Image reproduced by courtesy of Meade Instrument Corp.)

off-set guider
See *guide telescope.*

off-set guiding
A technique of *guiding*, used when the main object being imaged is very faint, or diffuse (e.g. a galaxy or interstellar nebula). The object is first centered in the main instrument. The *guide telescope* is then offset from the main telescope until it is aligned on a nearby bright star. Guiding then continues using that bright star. See also: *off-axis guider; off-axis guiding.*

Okayama observatory
An optical observatory at Okayama, Japan.

Old Moon

The phase of the Moon when it is in *conjunction* with the Sun, and so we see only its dark side. Also the thin *crescent* Moon for a few days before conjunction. See also: *Full Moon; New Moon; Half Moon; gibbous phase.*

Ondrejov observatory

An optical observatory with a 2m and several other instruments at Ondrejov near Prague.

Onsala observatory

A radio and millimeter wave observatory at Onsala in Sweden.

Ootacamund observatory

A radio observatory at Ootacamund in India with a 530m array that is located on a hillside so that it is aligned parallel with the Earth's rotational axis.

opacity

A measure of the ability of a semi-transparent medium to absorb radiation. It is given by the ratio of the intensity of the emergent radiation to that of the incident radiation. The reciprocal of opacity is *transmittance*. See also: *optical depth; optical density.*

opposition

A geometrical alignment for an object further from the Sun than the Earth, when the Sun, Earth and Object are in a straight line. This is the best time to observe an outer planet (Mars to Pluto) since the planet is then at its closest to us and we can see the whole of the visible disk (cf *greatest elongation*). See also: *aspect.*

optical axis

The main axis of symmetry of an optical system. Images are usually at their best on or close to the optical axis.

O

optical density

Optical *filters* and *photographic* images are often characterized by their optical density, D. This expresses the reduction in the intensity of a beam of radiation passing through the filter as a power of ten, with the power involved being the optical density. Thus an optical density of 1 corresponds to a reduction in intensity by a factor of $\times 10$ ($= 10^1$), an optical density of 2 corresponds to a reduction in intensity by a factor of $\times 100$ ($= 10^2$), an optical density of 3 corresponds to a reduction in intensity by a factor of $\times 1000$ ($= 10^3$), and so on. Shade numbers are also used to characterize filters. The shade number is given by 7D/3.

optical depth

A commonly used measure of the *opacity* of a semi-transparent medium, usually given the symbol, τ, and given by, $\tau = -\log_e$ (*Opacity*). An optical depth of 1 corresponds to the apparent surface of a semi-transparent object or region, such as the solar photosphere.

optical element

Any optically active component of an optical system such as a *lens*, *mirror*, *prism*, *diffraction grating*, etc., or the individual lenses within a *compound lens*.

optical flat

A mirror whose surface is flat to within an eighth of the operating wavelength. Used for example as the secondary mirror of a *Newtonian telescope*.

optical interferometer

See *interferometer.*

optical power

The reciprocal of the focal length (in meters) of a lens or less frequently, a mirror. Its units are dioptres. It is used when several lenses are combined in contact with each other to form an optical system. The power of the total system is then just the sum of the powers of the individual lenses.

optical quality

See *diffraction-limited optics.*

optical region

Electromagnetic radiation covering the *near infrared*, *visible* and *near ultraviolet* regions. The wavelength ranges from about 300 nm to about 5 μm. Over the optical region, telescopes, optical systems, detectors, etc. that are the same as or closely related to those for *visible light* may be used.

optical telescope

A telescope operating in *visible light*, or the *near infrared* or *near ultraviolet*. The main classes of optical telescope are the *refractor* and the *reflector* – see individual telescope designs for further details.

optical thickness

A synonym for *optical depth.*

optical transfer function

A synonym for *modulation transfer function.*

O

optical window

1 The range of medium wavelength *electromagnetic* waves that is able to penetrate to ground level on the Earth. It extends from about 300 nm to about 2 μm with additional small windows further into the infrared. See also: *atmospheric window; radio window.*
2 A plane piece of glass or other optical material whose surfaces are flat and parallel to an eighth of the operating wavelength, that is used to seal the entrance aperture to an instrument. See also: *Lanphier shutter.*

optics

Any optical component such as a *lens* or *mirror*. Also the science describing the passage of light through such optical components.

optimal filter

See *Wiener filter.*

optimum sampling

See *sampling theorem.*

orbital elements

Seven parameters that completely specify the properties of an orbiting body. They are:

- *semi-major axis* of the orbit (a)
- *eccentricity* of the orbital *ellipse* (e)

- inclination of the orbital plane to the *ecliptic* (i)
- orbital period (P)
- the time of perihelion passage (the epoch – T)
- longitude of the ascending node (Ω – the angle measured anti-clockwise around the ecliptic from the *First Point of Aries* to the ascending node – which is the intersection of the ecliptic and the orbital plane at which the body passes from south to north of the ecliptic)
- longitude of perihelion (ω – the angle, measured anticlockwise around the orbital plane from the ascending node to perihelion).

See also: *osculating elements.*

ordinary ray

See *birefringence.*

Orion Optics

A commercial firm manufacturing telescopes and accessories.

orrery

A mechanical model of a part or the whole of the solar system, mostly constructed in the seventeenth and eighteenth centuries.

A modern orrery showing the Earth's annual motion and the movement of the Moon around the Earth. (Image reproduced by courtesy of Carlo G. Croce, clockmaker.)

orrery software

A computer program that shows the positions and motions of planets and satellites within the solar system. It will often be a part of a *planetarium program.*

orthochromatic emulsion.

A *photographic emulsion* whose response mimics that of the human eye. See also: *panchromatic emulsion; orthochromatic emulsion.*

orthoscopic

An optical system that is completely corrected for all aberrations. See also: *achromatic; aplanatic; anastigmatic; apochromatic.*

orthoscopic eyepiece

A class B (see *eyepieces*) eyepiece made from an *achromatic* triplet and a *plano-convex lens.* It is practically free from all aberrations, has a useable *field of view* of about 50° and the *eye relief* is about the same as the focal length. The only drawback to the design is internal reflections producing *ghost images* but these can be reduced to negligible levels by *anti-reflection coatings* on the lens surfaces.

Oschin Schmidt camera
A 1.2m *Schmidt camera* at the *Mount Palomar observatory*.

oscillator strength
A measure of *spectrum line* strength. Except for hydrogen and hydrogen-like ions (atoms that have lost all but one electron), oscillator strengths have to be determined experimentally. The lack of this data, or its low accuracy when it is available, is a major constraint on our understanding of stellar spectra.

osculating elements
The instantaneously correct *orbital elements* for a body that has its orbit perturbed by the gravitational effect of another object. The actual position of the body will thus gradually diverge from that predicted from the osculating elements.

Osservatorio di Bologna
An optical observatory at Bologna, Italy.

Osservatorio di Brera
An observatory at Brera, Italy, that has been heavily involved with X-ray astronomy and its instrumentation.

over coating
See *aluminizing* and *mirror coating*.

over sampling
The use of a detector with a higher resolution than that required to obtain all the information that is present in an image. Over sampling does not cause any problems, but does represent a waste of money.

Owens valley observatory
A radio and millimeter wave observatory at Big Pine, California. It has several instruments including a 40m dish.

oxygen-III filter
See *nebular filter*.

pair production
The interaction that forms the basis of a number of detectors including *photography*, *CCDs* and *photodiodes*. In the process a substance, usually silicon, absorbs a photon. The energy from the photon lifts an electron from the valence to the conduction band, and leaves a positive "hole" in the valance band. The name of the process thus derives from the production of an electron–hole pair.

Palomar observatory
See *Mount Palomar observatory.*

Palomar observatory sky survey
A photographic survey of the sky covering from declination –30° to the north pole, obtained using the 1.2m *Oschin Schmidt camera* at *Mount Palomar.* A digitized version is available at http://dposs.caltech.edu/. See also: *STScI digitized sky survey.*

panchromatic emulsion.
A *photographic emulsion* with a more-or-less uniform sensitivity over the *visible region*. See also: *isochromatic emulsion; orthochromatic emulsion.*

Panoptic eyepiece
A modern class A (see *eyepieces*) eyepiece with six elements and a *field of view* of up to 68°. The *eye relief* is about two thirds of the focal length. The design is well corrected for aberrations.

parabola corrector
A *correcting lens* or lenses designed to reduce the *aberrations*, especially *coma*, produced in off-axis images by a *parabolic mirror.*

parabolic antenna
A radio *antenna* with a large *paraboloid* reflector that concentrates the radiation onto the *feed* and also helps to shield it from unwanted signals.

parabolic mirror
A mirror whose surface has the shape of a parabola (the shape produced by plotting $y = x^2$). Only parabolic mirrors are able to bring a parallel light beam to a sharp *focus*, provided that it is on the *optical axis* of the mirror. Spherical mirrors and simple lenses suffer from *spherical aberration*, and other image faults, as does the parabolic mirror when it produces images away from the optical axis. Most *reflectors* use parabolic mirrors as their *objectives*, although a variation on the *Cassegrain telescope* known as the *Ritchey–Chrétien* uses a hyperbolic mirror. See also: *conic section.*

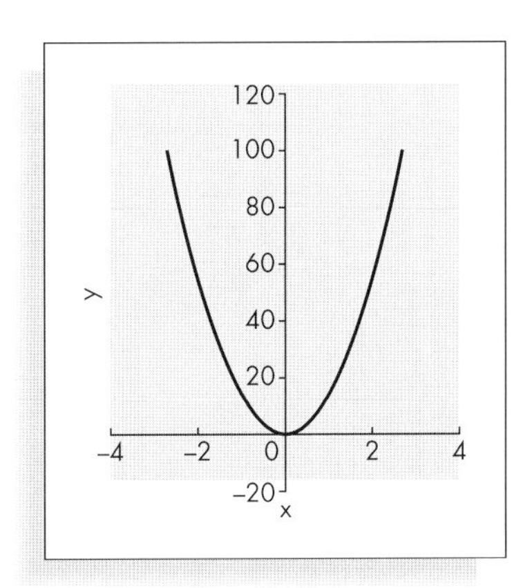

A parabola – most telescope parabolic mirrors are follow the part of the curve around the zero point, although glancing incidence mirrors come from high up the sides of the curve.

paraboloid

A shape obtained by rotating a parabola around its center line. Mirrors with paraboloidal shapes are used as the primary mirrors in *Newtonian and Cassegrain telescopes*, and in *grazing-incidence telescopes*. See also: *hyperboloid; conic section.*

parallax

The angle between the directions to a nearby object when observed from two different positions. In astronomy it usually refers to the change in position of a star as the Earth moves around its orbit. The parallax angle, P, (which is half the maximum movement of the star, and so equal to the angle subtended by the Earth's orbital radius as seen from the star) is related to the star's distance in parsecs by $D = P^{-1}$ when P is in seconds of arc. See also: *aberration (stellar); diurnal parallax.*

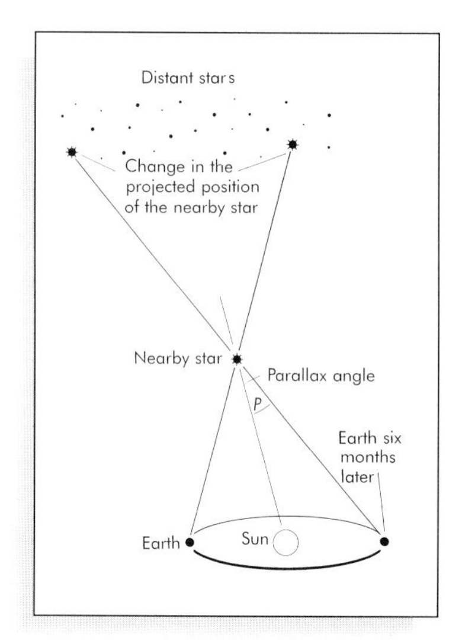

Stellar parallax due to the changing position of the Earth.

parametric amplifier

A low noise amplifier used in *radio receivers* at frequencies above 1 GHz.

parasitic antenna

See *Yagi antenna*. See also: *antenna; half-wave dipole.*

paraxial rays

Light rays that are always close to the *optical axis* of an optical system.

P

parfocal eyepieces

A set of *eyepieces* that have the same focus position. One such eyepiece may thus be exchanged for another without having to re-focus the telescope.

Parkes radio telescope

A 64m steerable radio dish at Parkes in Australia. See also: *Australia telescope.*

Parks Optical

A commercial firm manufacturing telescopes and accessories.

parsec

The distance at which a star has a *parallax* angle of 1″, symbol, pc. 1 pc = 3.09×10^{16} m = 3.26 *light years*. See also: *astronomical unit.*

Particle Physics and Astronomy Research Council

The main UK organization promoting professional astronomy, running observing facilities like the *Isaac Newton Group*, the *Gemini telescopes*, and *MERLIN*, and promoting public interest in astronomy.

Paschen series

See *Balmer series.*

passband
The range of wavelengths transmitted by a *filter*. See also: *bandwidth.*

pc
See *parsec.*

P Cygni line profile
A *spectrum line* with an emission component at the normal wavelength, and an absorption component shifted to shorter wavelengths. Such lines arise in gaseous envelopes around stars when the material in that envelope is expanding and accelerating outwards. Named for the archetypal star, P Cyg.

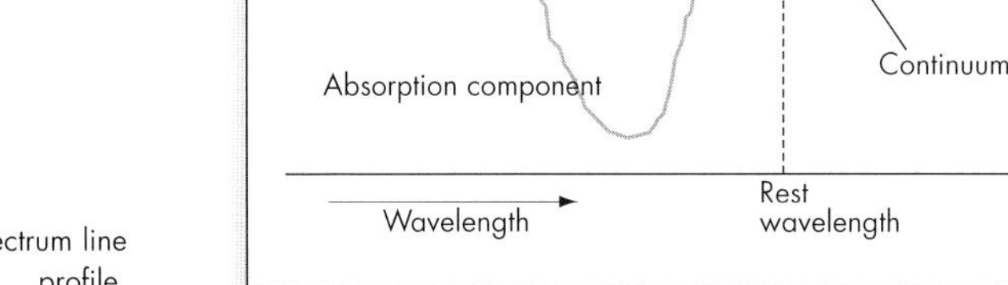

The P Cygni spectrum line profile.

Peltier cooler
A cooling device used for many small commercial *CCD* detectors whose action is based upon the Peltier effect.

pencil beam
A *main lobe* for a *radio telescope* that is narrow in both axes. See also: *fan beam.*

penta-prism
A five-sided prism that is used in single lens reflex cameras to produce an upright image. See also: *erecting prism.*

period
The time interval between repetitions of the same phenomenon during a cyclic change. It is the reciprocal of *frequency*. The period tends to be used as the measure for cycles that last for more than one second, while frequency tends to be used for cycles of less than one second duration.

periodic error correction
A system for correcting mechanical defects in the worm gear of a telescope drive that result in periodic errors in the telescope's *tracking* rate. On modern small telescopes, it requires the observer to *guide* accurately on an object while the computer control system for the drive learns the error corrections that are needed.

periodogram
A plot of the frequencies present in a signal (see power spectrum in *Fourier transforms*).

personal equation
An allowance made for personal bias in making a measurement. For example, the sunspot number is calculated using, $R = k\,(10g + s)$ where g is the number of groups, s is the number of individual spots. k is then the personal equation that allows for the efficiency of the observer and his/her equipment. It has to be determined by finding the spot number and comparing the observer's value with the official value at the same instant over a period of several months.

phase

1 A measure of the distance along a wave between two maxima starting from one of the maxima. The phase is usually given as an angle, based upon the idea of the wave originating via a circular motion. A maximum then has a phase angle of 0°. The angle increases smoothly along the wave, through 180° at a minimum to 360° (or 0° again) at the next maximum. Two waves are said to be in phase with each other if the difference between their phases at a point is 0° or an integer multiple of 360°. The waves are out of phase if the difference between their phases at a point is 180° or an integer multiple of 360° plus 180°. The former leads to constructive *interference*, and the latter to destructive interference.

2 The amount of the disk of the Moon or planet, etc. seen as illuminated from the Earth. (Tabulated annually in the Astronomical *Almanac*.)

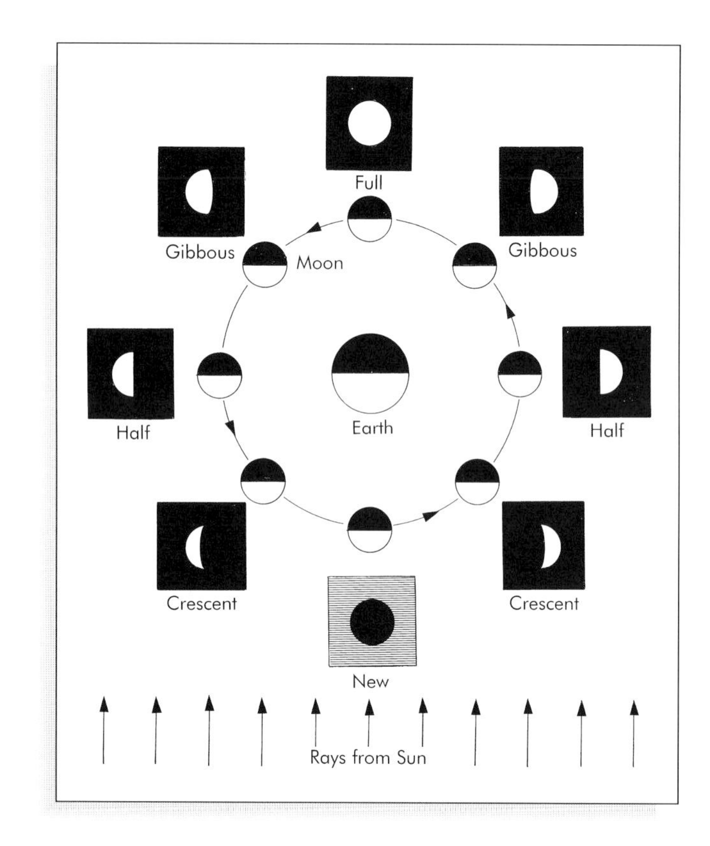

Phases of the Moon.

phase angle

The angular distance between the Sun and the Earth as seen from a third body within the solar system. See also: *aspect.*

phased array

A *radio telescope* that is formed from numerous simple fixed *antennas* set out in the form of a grid. It can be made very sensitive since a large area can be covered at low cost. The telescope can be "pointed" by steering its beam across the sky. The steering is accomplished by changing the *phase* delays between successive rows or columns of antennas in the grid. The 4-acre phased array at Cambridge made the discovery observation of pulsars in 1967.

phase plate

A synonym for a *wave plate.*

phase rotator

A component of an *aperture synthesis* system that maintains a zero *phase* difference as the object moves across the sky.

phase sensitive detector

An electronic device that detects periodic signals. It is able to extract weak signals from noise by *integrating* across many of the individual cycles of the signal. Signals that are not intrinsically periodic may be made so by *chopping.*

phase-switched interferometer

A radio *interferometer* in which the *phase* of the signal from one of the *antennas* is periodically changed by 180°. Switching an appropriate length of cable into and out of the transmission line is the usual method of changing the phase. The instrument has increased stability and sensitivity compared with a simple interferometer.

phoswich detector
See *scintillation detector.*

photo-cathode
The detecting element of a *photomultiplier*. It is coated with a photoemitter such as caesium antimonide.

photocell
See *photodiode.*

photoconductive cell
A detector used in the infrared region. Its response to the radiation is an increase in its conductivity as electrons are raised to the valence band. The change is detected by monitoring a small bias current through the device. Commonly used materials for photoconductive cells include indium antimonide and germanium doped with a variety of other elements. See also: *bolometer.*

photodiode
A detector operating in the optical region, formed by the junction between p-doped and n-doped areas of silicon. Photons absorbed in this region excite electrons to the conduction band where they flow from the p to the n region. This current is proportional to the intensity of the illumination and may then be measured. The size of the detecting area may be increased by inserting a layer of undoped (or intrinsic) silicon between the two doped regions. This then produces a p-i-n photodiode.

If a voltage of around 200 V is applied across the photodiode, then the original electron–hole pair are accelerated until they produce further pairs, and so on until perhaps a hundred or more electron–hole pairs result from a single detection. This is then called an avalanche photo-diode. See also: *pair production.*

photoelectric effect
The emission of electrons from the surface of a substance when it is illuminated by *photons* with sufficient energy. It forms the basis of a number of detectors including the *image intensifier* and the *photomultiplier*. See also: *work function.*

photoelectric magnitude
The *apparent magnitude* of an object measured using a photoelectric detector, especially a *photomultiplier*. See also: *UBV photometric system.*

photographic camera
A *camera* that uses *photographic film* or *plates*. See also: *single lens reflex camera.*

photographic contrast enhancement
A technique for enhancing the contrast of faint photographic images. In such images, the developed grains of the image lie predominantly in the top layer of the emulsion, while the background *fog* grains are distributed throughout the emulsion. A contact *print* is therefore made using diffuse light that preferentially reproduces the surface layer, thus increasing the *signal to noise ratio*. See also: *hypersensitization.*

photographic density
See *optical density.*

photographic emulsion

The main astronomical imaging detector for optical radiation used over the last one and a half centuries. It is now superseded for many purposes by the *charge-coupled device* (CCD), but still finds uses because of its cheapness and convenience. The emulsion is formed of tiny (1 μm) crystals of silver bromide with a small percentage of silver iodide in solid solution that are suspended in a matrix of gelatine. A photon absorbed by a crystal releases an electron (see *pair production*) that converts a silver ion to a silver atom. Between 10 and 20 silver atoms must be produced in a single crystal before they stabilize. This cluster of silver atoms is called the *latent image*. After exposure the emulsion must be *developed*. This is a chemical process that converts silver bromide to silver. It is, however, catalyzed by silver atoms. The crystals containing the silver atoms of the latent image thus develop (convert entirely to silver) more rapidly than any pure silver bromide crystals. After developing, the emulsion is *fixed* in another chemical process that dissolves away the remaining silver bromide. The photographic emulsion then contains the required image, with the greatest density of silver particles corresponding to the most intense areas of illumination. Since the silver particles absorb light, the photographic image is thus a *negative* one. To obtain a normal image the negative has to be re-photographed, a process called *printing*, however, for astronomical purposes, the negatives are usually worked on directly.

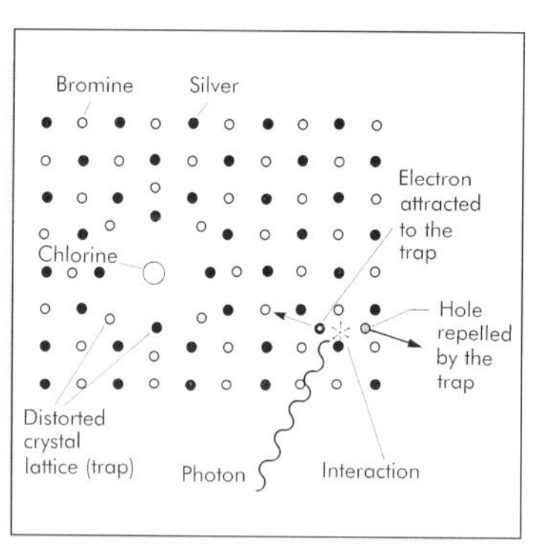

An electron trap within a photographic emulsion formed by a chlorine atom within a silver bromide grain.

Photographic emulsion has a low *quantum efficiency* (0.1% to 1%) partly because many photons have to be absorbed within each crystal before a stable latent image is formed. The speed of response may be varied somewhat by changing the size of the silver bromide crystals. A faster emulsion has larger crystals. The larger crystals, however, produce a coarser (lower spatial resolution) emulsion, and so a compromise has to be reached between speed and resolution. At long exposures, the speed of an emulsion reduces markedly, a phenomenon called *reciprocity failure*, because the reciprocal relationship between image intensity and required exposure length has broken down. Reciprocity failure is minimized, but not eliminated, in some emulsions produced specifically for astronomical use.

Basic photographic emulsion responds to wavelengths from about 450 nm to 350 nm, but this may be extended to longer wavelengths by adding dyes to the crystals. The short wavelength limit arises because of absorption in the gelatine, and so may be extended by using a different or no matrix. By these means, photography may be used over the range 1.1 μm to about 300 nm. The image produce on a single type of emulsion is black and white (sometimes called monochromatic). Combining emulsions with different spectral responses produces *color film*. See also: *fog; photographic speed; panchromatic emulsion; orthochromatic emulsion; isochromatic emulsion.*

photographic film

The most commonly available material for *photography*. It is a thin layer of *photographic emulsion* coated onto a flexible plastic film.

photographic magnitude

The *apparent magnitude* of an object determined from an image on a blue-sensitive *photographic emulsion.*

photographic plate

A thin layer of *photographic emulsion* coated onto a thin sheet of glass. For astronomical use it is to be preferred over *photographic film* when precise positional measurements are to be made or the material is to be archived, since it is less subject to distortion. Photographic plates, however, are less widely available than film, and may have to be ordered specially from the suppliers.

photographic speed

A measure of the *exposure* required to produce an image on a *photographic emulsion.* Faster emulsions require shorter exposures. Its ISO number shows the speed of an emulsion. This is a combination of two earlier systems; the arithmetic ASA scale and the logarithmic DIN scale. The first number in the ISO scale is the ASA number, and the second is the DIN number. The required exposure will be inversely proportional to the ASA number, and will halve for every three-unit increase in the DIN number.

Slow emulsions are around ISO 50/18°, medium speed emulsions range from ISO 100/21° to ISO 200/24°, while fast emulsions range from ISO 400/27° to ISO 1600/34°.

Since faster emulsions have larger crystals of silver bromide, the images they produce will be coarser and less able to withstand enlarging. See also: *hypersensitization.*

photographic zenith tube

A vertically pointing telescope that uses a bath of mercury to define the zenith exactly. It is used to calibrate clocks by observing stars close to the zenith. See also: *zenith telescope.*

photography

The process of obtaining images using *photographic emulsion*. The photographic *plate* or *film* is often held inside a *camera*, but for astronomical purposes, the telescope replaces the camera lens, and so all that is needed is a plate or film holder at the telescope focus together with a movable cover to start and stop the exposure.

photometer

An instrument designed for the purpose of measuring the intensity of electromagnetic radiation emitted by an object in the sky. In the *optical region*, photometers are normally based upon *photomultipliers*, although devices based upon *photo diodes* are manufactured for use on small telescopes. A photometer normally has an entrance aperture to cut down background light, a viewing system for aligning onto the object and several interchangeable filters. The filters are usually for one of the standard photometric systems (see *UBV photometric system* and *Strömgren system*). For photomultiplier-based instruments, the use of a *Fabry lens* will help to reduce the intrinsic *noise* due to variations in the sensitivity across the photocathode. *Photography* and more especially *CCD* detectors are able to provide information on the intensity of objects, and the latter is replacing purpose built photometers for many purposes. See also: *photometry; radiometer.*

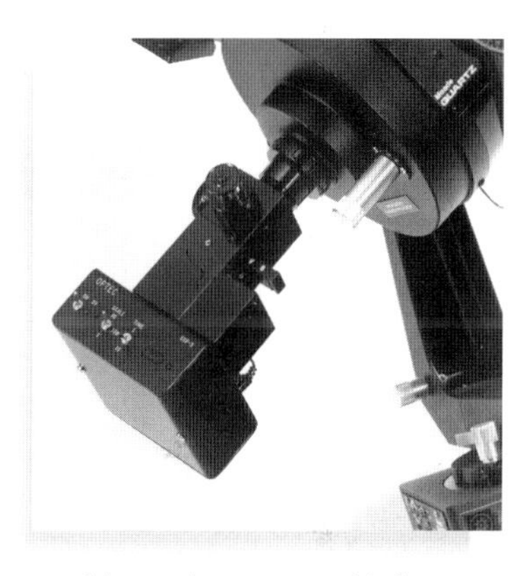

A solid-state photometer suitable for use on small telescopes. (Image reproduced by courtesy of Optec Inc.)

photometer, multi-channel

See *multi-channel photometer.*

photometer, two star

See *two star photometer.*

photometric system

See *UBV photometric system* and *Strömgren system.*

photometry

The science of measuring the intensity of electromagnetic radiation emitted by an object in the sky. Photometry occurs at all wavelengths, but the term is perhaps most often used in connection with *optical* and *infrared* measurements. The *wavelength* and *bandwidth* of the measurements are usually defined and restricted by the use of *filters* (see *UBV photometric system*), although at some infrared wavelengths it is simply the available *atmospheric window* that is used. Accurate photometry demands excellent observing conditions and so is best carried out from high altitude observatories. See also: *photometer; apparent magnitude; color index.*

photomultiplier

A detector operating over the *optical region* and into the *ultraviolet*. The device has a high intrinsic amplification and is often used in *photometers* and for *photon counting*. A photomultiplier consists of an evacuated glass tube with a window transparent to the required radiation. Behind the window is the *photo-cathode*. This is an electrode at a potential of about −2000 V that has a photo-emitting substance, such as caesium antimonide, coated onto its surface. At the other end of the tube is the anode, and a number of subsidiary electrodes, called dynodes, are strung out between the photo-cathode and the anode. The dynodes are at lower voltages that decrease from the photo-cathode voltage by about 100 V to 200 V for each successive dynode. The dynodes are also coated, again usually with caesium antimonide. An incoming photon, produces a photoelectron when it is absorbed in the coating on the photo-cathode. This electron is attracted to the first dynode by the lower negative (i.e. more positive) voltage. The potential difference between the photo-cathode and the first dynode accelerates the photoelectron so that when it hits the first dynode many secondary electrons are produced. Between 5 and 20 secondary electrons are produced from the single incoming photoelectron. Those secondary electrons are then accelerated to the second dynode, where each one produces another 5 to 20 electrons. This process repeats down the length of the photomultiplier, until the anode gathers 10^6 or more electrons for a single incoming photon.

The photomultiplier has a high *quantum efficiency* and a high intrinsic amplification. Its photo-cathode may be coated with a variety of photo-emitters to give responses from 1.1 μm to 150 nm. But it also has a number of drawbacks that are leading to its replacement by *CCDs* for some purposes. Those drawbacks include the high voltages required, the fragile nature of the device, a variation in the response depending whereabouts the photon hits the photo-cathode, and a degree of *non-linearity* in its response.

photon

See *electro-magnetic radiation*.

photon counting

An observing technique that leads to reduced noise when detecting faint signals. The detector is usually a *photomultiplier* or another detector that is sufficiently sensitive to give a detectable signal for a single incoming photon. The intensity of the source is then measured by counting the number of photons detected, rather than by the magnitude of the output current. This leads to a reduction in the noise, because the variable response of a photomultiplier means that two identical photons may produce slightly different output currents, but each pulse counts as one detection whatever its size may be. A further reduction in noise may be achieved by putting in thresholds below or above which the output pulse is not counted. Very low pulses probably originate from thermal electrons within the photomultiplier, while very strong ones arise from Čerenkov radiation as a cosmic ray passes through the photomultiplier's window.

photon noise

See *quantum noise*.

photopic vision

Vision under bright levels illumination, arising from the cone cells in the *retina*. See also: *scotopic vision*.

photovisual magnitude

The *apparent magnitude* of an object determined from an image on a *photographic emulsion* whose response mimics that of the eye (e.g. *orthochromatic emulsion*).

photovoltaic cell
See *photodiode.*

Pic du Midi observatory
An optical observatory sited in the French Pyrenees at an altitude of 2900 m. It houses a 2m telescope amongst several others.

Pickering lines
The *spectrum lines* of the singly *ionized* helium atom in the *visible* part of the spectrum. Since this ion is physically similar to the hydrogen atom, the Pickering series is similar in appearance to the *Balmer series,* but with an extra line between each Balmer line. The increased nuclear charge means that these visible lines arise from transitions to and from the third excited level rather than the first. The first few lines in the series are at 1012.4 nm, 656.0 nm, 541.2 nm, 485.9 nm and 454.1 nm.

pier
A support for a small *telescope mounting* comprising a substantial vertical pillar usually of round or square cross section. Permanent piers may be made of cast concrete and have foundations a meter or more deep in order to be as vibration-free as possible. Portable piers may take the form of a central pillar with three or four horizontal buttresses that may be disassembled for transportation. See also: *alt-azimuth telescope mounting; equatorial telescope mounting; tripod.*

piggy-back mount
A mounting bracket, etc. that is attached to a telescope and that enables other instrumentation such as a camera, a second telescope, etc. to be attached to the main instrument. The *tracking* of the main telescope then allows the ancillary instrument to follow the object in the sky.

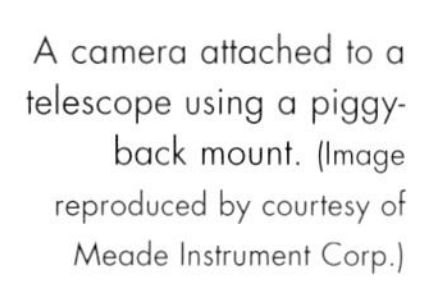

A camera attached to a telescope using a piggy-back mount. (Image reproduced by courtesy of Meade Instrument Corp.)

pin cushion distortion
See *distortion.*

pin-hole camera
A camera that does not have a lens. It has a small entrance aperture (a pin-hole) and an upside-down image is formed by the light rays simply passing straight through the aperture. The resolution of the device is determined by the size of the pinhole, but the smaller the pinhole the less energy reaches the image. None-the-less some X-ray and gamma ray detectors operate essentially as pinhole cameras.

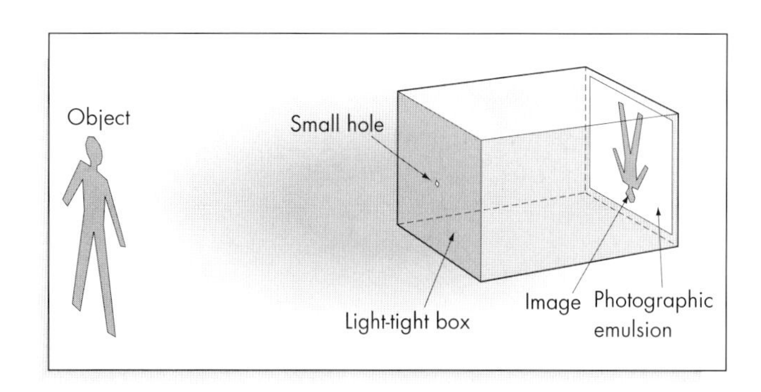

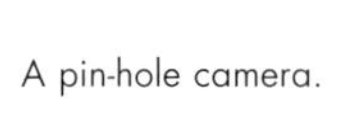

A pin-hole camera.

p-i-n photodiode
See *photodiode.*

pitch lap
A covering of pitch onto the tool that is used in conjunction with a polishing agent to *polish* and *figure* a mirror. See also: *grinding.*

pixel
An individual detecting element in imaging detectors formed from a grid of such elements (such as a *CCD detector*), or the corresponding bit of an image produced by such a detector (the name is a twisted concatenation of **pic**ture **el**ement).

plane polarization
A synonym for *linear polarization.*

planetarium
A mechanism for simulating the night sky, also the building that houses such a mechanism. The planetarium usually has a number of projectors that produce pinpoints of light (stars) or images of the Moon and planets on the inside of a hemispherical dome and which can be rotated and moved to simulate diurnal and other motions of the sky. The audience sits within the dome to view the scene.

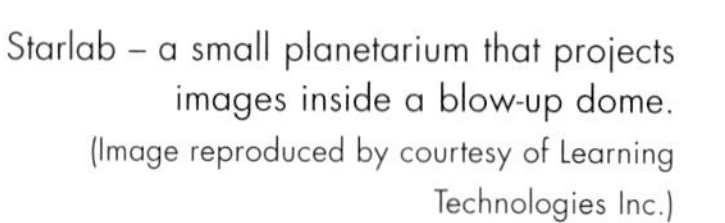
Starlab – a small planetarium that projects images inside a blow-up dome.
(Image reproduced by courtesy of Learning Technologies Inc.)

P

planetarium program
A computer program that displays views of the sky for a particular place and time. It will usually also include the positions of any of the planets that may be visible and the phase of the Moon. See also: *orrery software.*

planetary filter
A colored filter used to enhance observation of the markings on planets. For example, the dark features on Mars will have an improved contrast through a yellow or red filter, while blue or green filters will show the clouds and polar caps more clearly.

planetary precession
A small change in the position in space of the Earth's orbital plane due to *precession* caused by gravitational perturbations from the planets. Its period is about 13 million years. See also: *precession of the equinoxes.*

Planetary Society
A society based in Pasadena, California, but international in scope, promoting interest in space activities. See http://planetary.org.

planisphere

A device that shows the stars visible from a particular latitude at any time of the night or year. It comprises two circular sheets pivoted at their centers. The bottom sheet has a map of all parts of the sky visible from the latitude. The top sheet has an elliptical window that represents the hemisphere of the sky visible at a particular instant. Scales on the edges of the sheets allow the top one to be rotated to show the stars visible at a particular instant.

Planispheres.

plano-concave lens

A *lens* with a concave surface on one side and a flat one on the other. See also: *diverging lens.*

plano-convex lens

A *lens* with a convex surface on one side and a flat one on the other. See also: *converging lens.*

plasma

A wholly or partially ionized gas. Stars, H II regions, planetary nebulae, supernova remnants, etc., and much of the interstellar medium are formed from plasmas. The free electrons within a plasma give it a very high conductivity and so it interacts strongly with magnetic fields. The movement of the plasma then either drags the magnetic field or the plasma is constrained to follow the direction of the magnetic field.

plasma frequency

The natural frequency of oscillation within a plasma. It is given by:

$$f_p = \sqrt{\frac{e^2 N_e}{4\pi^2 \varepsilon_o m_e}} \approx 8.98 N_e^{1/2} \text{ Hz},$$

where e is the charge on the electron (1.602×10^{-19} C), N_e is the number of electrons per m^3, ε_o is the permittivity of a vacuum (8.85×10^{-12} F m^{-1}) and m_e is the electron mass (9.110×10^{-31} kg). Electromagnetic radiation with a higher frequency than the plasma frequency can propagate through the material, but lower frequencies will be absorbed or reflected. The Earth's ionosphere thus reflects frequencies less than about 30 MHz, allowing long distance radio communication to take place, but also limits radio astronomy to frequencies higher than about 10 MHz.

plate constants

The factors to convert a linear measurement of the position of an image on a photographic plate to its position in the sky.

plate measuring machine

A machine for measuring the positions of images on a photographic plate. It is essentially a form of travelling microscope. The plate is placed on a table that can be moved in one or two directions and that has accurate read-outs of its position. The positions of objects are determined by centring them in the field of view of the microscope. Nowadays automatic computer controled machines largely undertake the measurements. See also: *microdensitometer.*

plate scale

See *image scale.*

Platonic year

The period of the *precession of the equinoxes*, equal to about 25,800 years.

Plössl eyepiece

A class B (see *eyepieces*) eyepiece that is an off-shoot of the *Kellner*. It is made from two closely spaced *achromatic lenses* and is virtually *orthoscopic* (free from all *aberrations*). It has a useable *field of view* of up to 55°, and the *eye relief* is about 80% of the focal length. A fifth lens may sometimes be added between the two achromats to produce a super-Plössl . The design can be used on relatively fast telescopes (f5 or faster), and is the eyepiece of choice for many observers.

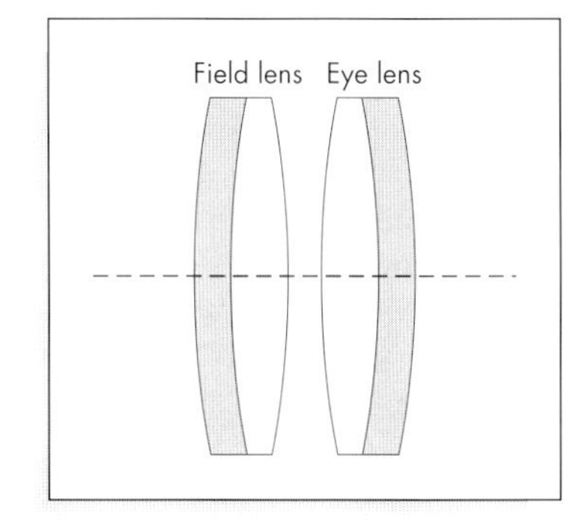

The optical arrangement of the Plössl eyepiece.

Pockel's cell

A switchable *wave plate* for use in *polarizers* whose operating principle is based upon the Pockel's effect. The Pockel's effect is the induction of birefringence into a material by the application of an electric field. Using materials such as ammonium dihydrogen phosphate Pockel's cells can be made that can be switched between acting as *quarter* and *half wave plates* at up to 10^6 cycles per second.

Pogson's equation

The relationship between the brightnesses of objects and their *apparent magnitudes* given by:

$$m_1 - m_2 = -2.5 \log_{10}\left(\frac{B_1}{B_2}\right),$$

where m_1 and m_2 are the magnitudes of the two objects, and B_1 and B_2 are their brightnesses or energies.

Pogson step method

A variant of the *fractional estimate* method of obtaining visual apparent magnitudes, using only a single comparison star. See also: *Argelander step method.*

point source

An object whose image is smaller than the *resolution* of the telescope and/or which is contained within one *pixel* of the detector. See also: *extended source.*

point spread function

The result of an instrument observing a point source or its equivalent (e.g. a monochromatic emission line for a *spectroscope*). It is also called the instrumental profile and often abbreviated to PSF. Because of *diffraction*, scattering, *aberrations*, etc. the image of a point source is not itself a point. The point spread function will usually have a central peak into which 50% to 80% of the energy is concentrated with the remaining energy spread into extensive wings that may have diffraction fringes superimposed on them. The point-spread function determines the *resolution* and *modulation transfer function* of the instrument and is used in image processing procedures such as *CLEAN* and *maximum entropy method*. See also: *Airy disk; main lobe; side lobe.*

polar axis

On an *equatorial telescope mounting*, the axis, movement around which changes the telescope's *right ascension*. The polar axis is parallel to the Earth's rotational axis. The telescope can therefore *track* objects through a simple constant velocity rotation around the polar axis at one revolution per *sidereal day*. The polar axis is orthogonal to the *declination axis*. See also: *alignment*.

polar diagram

A plot in circular coordinates of the way that the sensitivity of a *radio antenna* or *telescope* varies with angle. In most cases there will be a peak in the sensitivity given by the *main lobe*. Often there will also be some sensitivity at various angles to the main lobe, shown by the *side lobes*. The main lobe and side lobes are the equivalent of the *Airy disk* and surrounding fringes of an optical telescope.

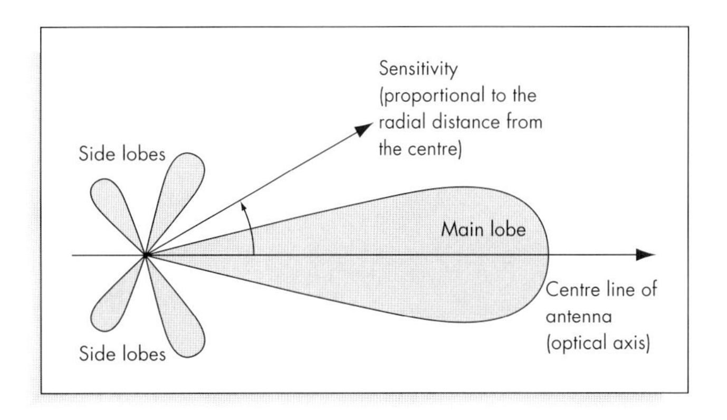

A schematic polar diagram for a radio antenna.

polar distance

The angular distance of an object in the sky from the north *celestial pole*. It is equal to 90° – *declination*.

polar flattening

A synonym for *oblateness*.

polarimeter

An instrument for measuring the degree and type of *polarization* present in the radiation from astronomical objects. In the *optical region* such instruments are essentially a photometer that has a *polarizer* incorporated into the optical path. The varying output as the polarizer is rotated then provides the measure of the polarization of the light from the object. In the *radio region*, many *antennas* are intrinsically sensitive to the polarization of the radiation.

polarimetry

The science of studying the *polarization* of *electromagnetic radiation* emitted by objects in the sky.

polarization

Any degree of non-randomness in the distribution of a quantity. Most commonly used for *electromagnetic radiation*. See also: *polarized radiation*.

polarized radiation

Electromagnetic radiation in which the direction of the electric vector is non-random. See also: *circularly polarized radiation; elliptically polarized radiation; linearly polarized radiation*.

polarizer

A device that only allows passage through it of *electromagnetic radiation* that is *linearly polarized* in some specified direction. It may be used to reduce *background light*

P

from the Moon, since this is strongly polarized for directions around 60° to 120° away from the Moon. The polarizer should be rotated whilst viewing through the eyepiece until the background light is minimized. It may also be used to highlight features of strongly polarized objects such as the Crab nebula. Two polarizers may be superimposed and one rotated to produce a variable density filter. However, such a filter is NOT suitable for solar observing (see *solar filter* for further details). See also: *Nicol prism; Polaroid.*

polarising filter

1 see *variable density filter.*
2 A filter comprising a single sheet of *Polaroid* that is used to reduce sky background. Scattered light from the Moon (or Sun) is partially *polarized* and at an angle of 90° to the Moon it is nearly 100% linearly polarized. A significant reduction in the sky background on moonlit nights can therefore be achieved for objects some distance away from the Moon, through the use of such a filter. It must be rotated until the position is found where the sky background is reduced the most.

Polaroid

A commercially produced *polarizer.* It is in the form of a thin sheet of plastic that contains many aligned *dichroic crystals.*

pole of the ecliptic

See *ecliptic pole.*

polishing

The final stage in the production of an optical surface, and which includes *figuring.* Polishing removes the last pits left by the *grinding* process. It is usually accomplished with a powder such as iron oxide or cerium oxide that is softer than the carborundum used for grinding. The powder is held on a matrix of soft material, such as pitch, that has the same shape as the optical surface. The matrix is first moved back and forth in a random pattern to polish the surface, and then in a regular pattern aimed at reducing high points to the required profile (figuring). See also: *lap.*

P

Poncet platform

A device that allows an *alt–azimuth-mounted* telescope (especially a *Dobsonian*) to track objects using a single, constant velocity motor. The whole telescope is mounted onto a horizontal platform, that in turn moves on inclined bearings aligned on the pole. See also: *equatorial platform.*

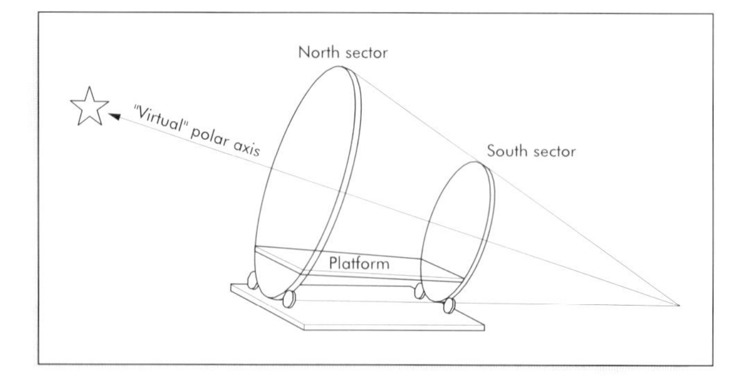

The principle of the Poncet platform.

population index

A measure of the rate at which the number of meteors increases as their brightness decreases. It is roughly 250% for each decrease in the brightness level by one stellar magnitude.

Porro prisms
A combination of two isosceles prisms each with a 90° angle and two 45° angles often used in *binoculars* to produce an upright image. See also: *erecting prism.*

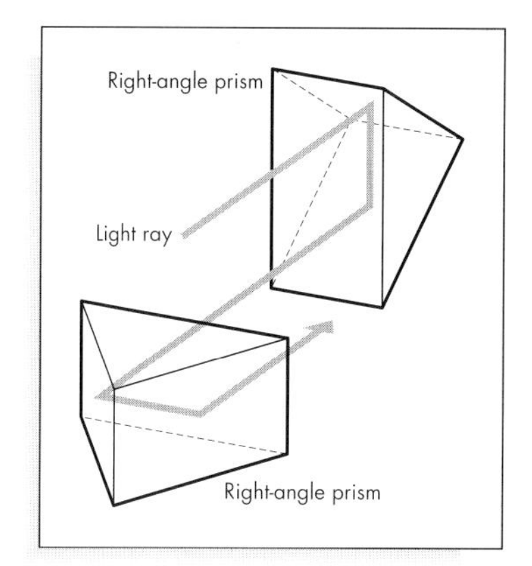

Light paths through a pair of Porro prisms.

Porter's solar disks
A synonym for *Stoneyhurst solar disks.*

position angle
The direction on the sky of one object relative to another. The position angle is the angle from the north in the sense N–E–S–W. See also: *double star; separation.*

position angle effect
A variation in the perceived relative brightnesses of two objects depending upon their relative orientations in the field of view. It arises from variations in the sensitivity of the eye's retina. It is a problem when making eye estimates of stellar magnitudes. See also: *Purkinje effect.*

position-sensitive proportional counter
A detector for X-rays, gamma rays and cosmic rays. It is a *proportional counter* that has a resistive central anode. The strength of the pulse generated by the high-energy photon or particle is measured at both ends of the anode. Since the pulse magnitudes will depend upon the distance along the anode to where the electron avalanche arrived, the position of that point can be determined. A grid of anodes can be used to produce an imaging detector.

positive image
An image wherein the brightest parts of the original object are reproduced as the brightest parts of the image, and the darkest parts of the original object are reproduced as the darkest parts of the image. Most images are naturally positive, but *photography* produces a *negative* original image that then has to be re-photographed to obtain a positive image. See also: *photographic emulsion; printing.*

positive lens
See *converging lens.*

power
A measure of the *magnification* provided by an *eyepiece*. Eyepieces for astronomical use are usually marked with their *focal length* that divides directly into the focal length of the *objective* to give the magnification. Eyepieces produced for use on microscopes, etc., however, will normally be marked with a magnification or power such as ×10. This is based upon a standard distance of 0.25 m for the image produced by the microscope objective. A ×10 eyepiece thus has a focal length of 25 mm, and a ×25 eyepiece, a focal length of 10 mm.

power spectrum
See *Fourier transform.*

PPARC

See *Particle Physics and Astronomy Research Council.*

pre-amplifier

The first stage of amplification within a *radio telescope*. It must have as low an intrinsic *noise* as possible, because its own noise will be amplified in later stages of the receiver. It is thus usually placed very close to the *feed*, and often cooled.

preceding

See *leading edge.*

precession

A change in the direction of the rotation axis of a rotating object arising from a twisting force. The rotational axis moves at right angles to the direction of the torque. The effect may easily be observed with a gyroscope or spinning top. The gravitational perturbations of the Sun and Moon on the rotating Earth, cause its axis to revolve around the *pole of the ecliptic* in Draco every 25,800 years, an effect called luni-solar precession (see *precession of the equinoxes*). Since an orbiting object may be regarded as rotating around the central object, precession can also lead to the plane of an orbit rotating in space. This effect causes the plane of the Moon's orbit to rotate through 360° every 18.61 years. See also: *planetary precession.*

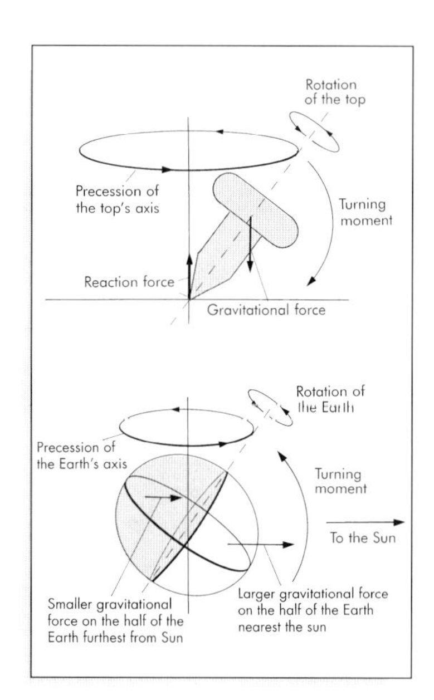

Precession for a spinning top and the Earth.

P

precession of the equinoxes

See panel (page 191).

primary aberration

See *aberration* (optical).

primary cosmic ray

See *cosmic ray.*

primary mirror

The *objective* of a *reflecting telescope.* Its function is to gather the light and to bring it to a *focus.* Its shape is *parabolic* in most telescope designs, but it is spherical in *Schmidt cameras,* and hyperbolic in *Ritchey–Chrétien telescopes.*

primary star

See *double star.*

prime focus

See panel (page 192).

precession of the equinoxes

The movement of the Earth's rotational axis arising from the *precession* caused by the torque from the Sun and Moon acting on the Earth's equatorial bulge. Since the equator is the plane orthogonal to the rotational axis, the polar motion causes the equator to move as well. This in turn leads to a change in the coordinates of objects in the sky, since *right ascension* and *declination* are measured with respect to the equator and its intersection (the *First point of Aries*) with the *ecliptic*. The First Point of Aries moves backwards around the ecliptic once in 25,800 years, a yearly motion of 50.3″. The effect of this on RA and Dec must be corrected if accurate positions of objects are needed. The corrections are given by:

$\Delta\alpha = 3.4\Delta T(\cos\epsilon + \sin\epsilon \sin a \tan\delta)$ seconds of time
$\Delta\delta = 50.3\Delta T \sin\epsilon \cos\alpha$ seconds of arc

where $\Delta\alpha$ is the change in RA, $\Delta\delta$ is the change in Dec, ΔT is the time interval in years between the *epoch* of the known position and the date for which the new position is needed, ϵ is the *obliquity of the ecliptic*, α is the right ascension (note that it must be converted to degrees before inserting into these formulae) and δ is the declination of the object. See also: *Platonic year.*

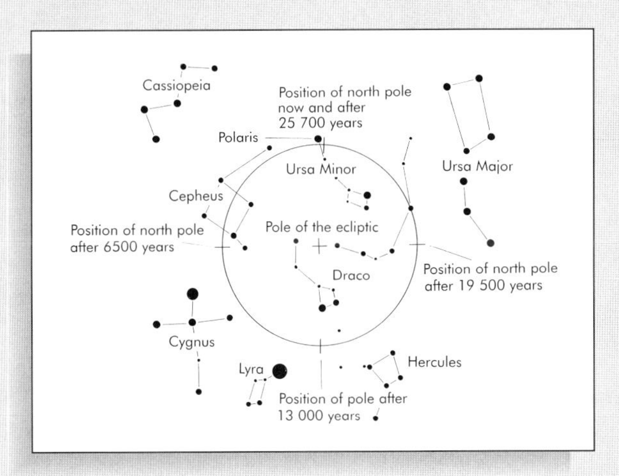

The movement of the North Celestial pole against the stars due to the precession of the equinoxes.

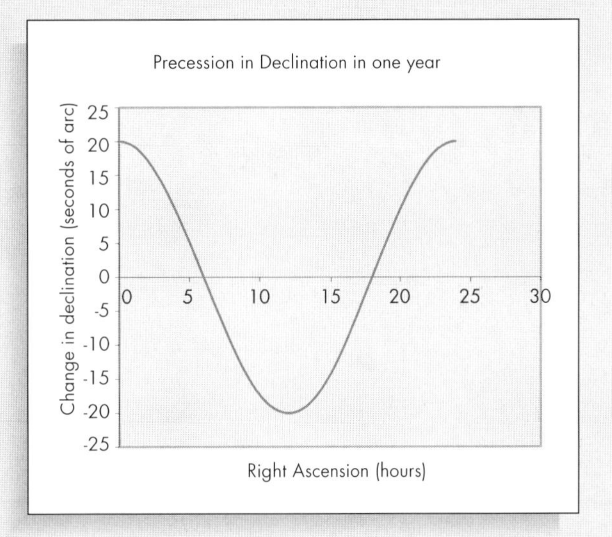

The annual precession in declination

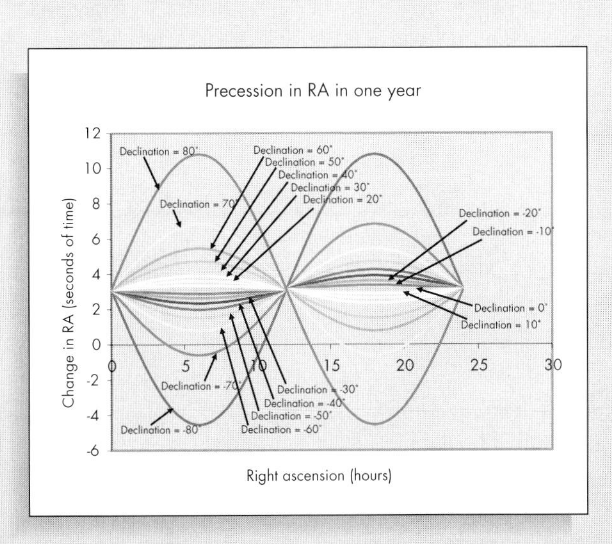

The annual precession in right ascension

prime focus

The *focus* of the *objective* of the telescope. This is the same as the focus of the telescope in most *refractors*. For *Newtonian telescopes* the secondary flat mirror reflects the focus to the side of the telescope, but the optical properties (*focal length*, *focal ratio*, *image scale*) are the same as at the prime focus. But in *Cassegrain telescopes* and similar designs, the secondary mirror acts to increase the focal ratio, and so the telescope focus has a larger image scale than the prime focus. Large telescopes are often used at prime focus, but *correcting lenses* are needed to give reasonable fields of view.

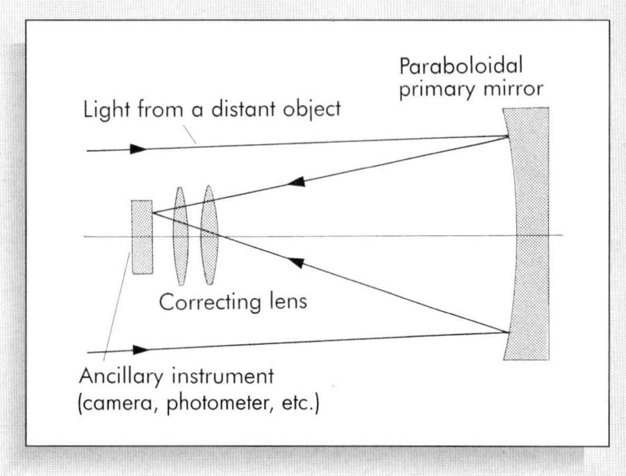

The prime focus position for a telescope.

The 5 m Hale telescope – the prime focus position is at the top of the tube. (Copyright A. Maury.)

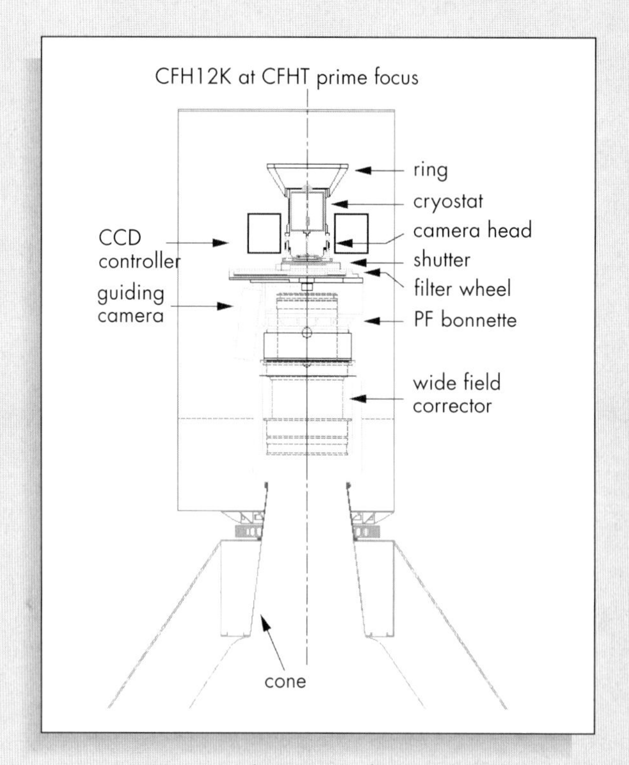

Instrumentation at the prime focus of the 3.6 m Canada-France-Hawaii telescope (Image reproduced by courtesy of the Canada-France-Hawaii observatory (1999) and J-C Cuillandre).

P

prime meridian

The *meridian* that passes through the zenith for an observer. It divides the sky into its eastern and western hemispheres. The *hour angle* of an object is then measured westwards from the prime meridian.

printing

The process in *photography* of re-photographing the original *negative image* in order to obtain a *positive image*. For precise work, the original negative image is to be preferred since the printing process inevitably introduces changes and distortions.

prism

A cylindrical object with a fixed cross-sectional shape. The usual cross-section is either an equilateral triangle or an isosceles triangle with a 90° angle and two 45° angles. The former type is used in *spectroscopes* as the dispersing element. The latter is used to reflect light through 90° or 180°. To reflect light through 90°, the light beam enters the prism through one of its short faces, is internally reflected off the long face, and emerges through the other short face. For an 180° change, the light enters perpendicularly to the long face of the prism and is internally reflected twice off the other two faces before re-emerging from the long face parallel to its original direction. See also: *Porro prism; Nicol prism; Wollaston prism.*

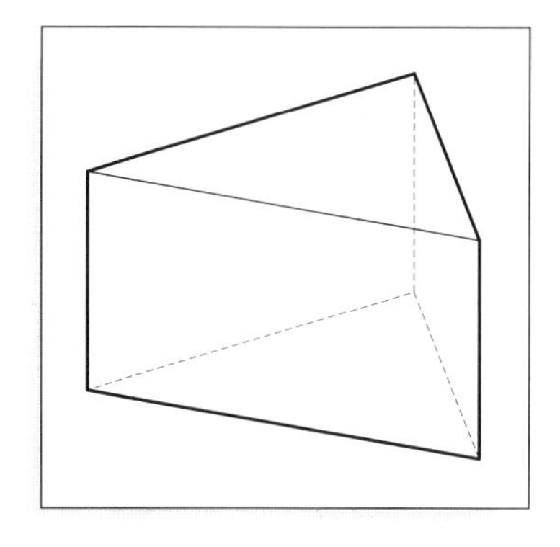

A 60° prism.

prismatic astrolabe

See *Danjon astrolabe.*

prism spectroscope

A *spectroscope* based upon a *prism*.

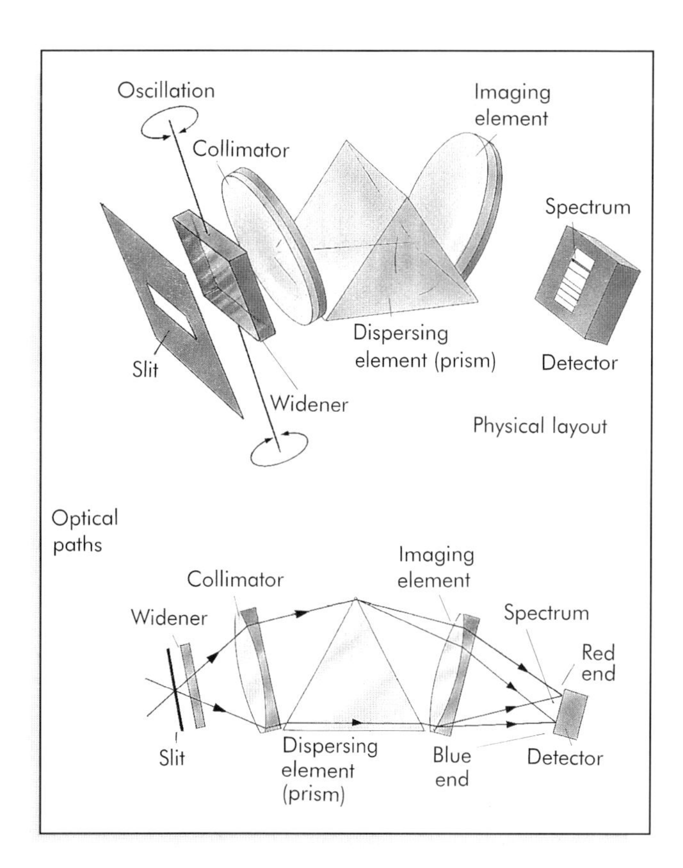

A prism-based spectroscope.

probable error

A measure of the *uncertainty* of a set of measurements. An individual measurement has an equal chance of being closer to the mean value of the set than the probable error or further from it. The probable error is given by, 0.6745 σ, where σ is the *standard deviation.*

prograde motion

See *direct motion;* See also: *retrograde motion.*

prominence spectroscope

An instrument for observing solar prominences. It is a *spectroscope* with an additional slit centered on one of the strong solar absorption lines, usually H-α. The entrance slit to the spectroscope is aligned on the limb of the Sun, and the observer looks through the second slit (CAUTION – always take appropriate precautions when observing the Sun or damage to the eye and/or telescope can result). Prominences may be then be seen because the prominence has an emission line spectrum and so emits strongly at that wavelength, while the solar photospheric emission is spread over the whole spectrum. See also: *spectrohelioscope; H-α filter.*

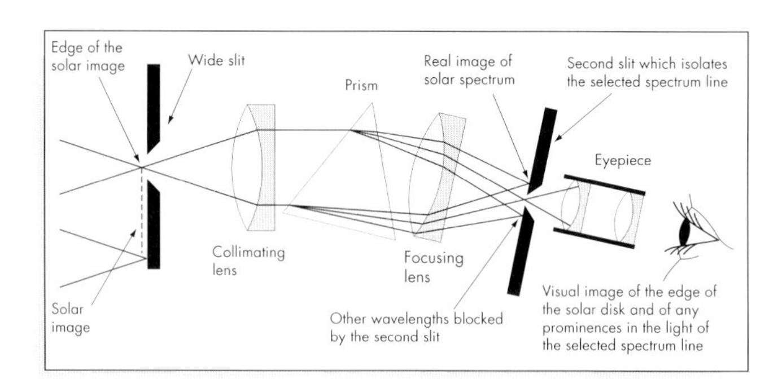

Light paths within a prominence spectroscope.

promscope

A synonym for a *prominence spectroscope.* The term is also sometimes used for the *H-α prominence telescope.*

Pro-optic

A commercial firm manufacturing telescopes and accessories.

proper motion

The movement of a star across the sky arising from its velocity through space relative to the Earth. Even for nearby stars it is usually only a fraction of a second of arc per year, and it is too small to measure for more distant stars. The second closest star to the Earth, Barnard's star in Ophiuchus, has the largest proper motion at 10.3″/year.

P

proportional counter

A detector for X-rays, gamma rays and cosmic rays. It is essentially a Geiger counter operated at a lower than normal voltage. It is an enclosure containing a gas such as argon that has a voltage applied between the walls of the enclosure and an anode running down the center. The ionising radiation produces electrons in the gas that are then accelerated towards the anode, and cause further ionizations on the way. A burst of electrons therefore arrives at the anode for every high-energy photon or cosmic ray entering the device. Unlike a Geiger counter, however, the electron avalanche does not saturate, and so the pulse strength is proportional to the energy of the photon or particle.

In the gas scintillation proportional counter, scintillations produced by the electrons within the gas are detected rather than the current pulses. See also: *position-sensitive proportional counter.*

proton

A sub-atomic particle that is a fundamental component of the nucleus of an atom. It has a mass of 1.673×10^{-27} kg, and a positive charge of 1.602×10^{-19} C. The number of protons in the nucleus is the atomic number and determines the element formed by the atom. See also: *electron; neutron.*

pseudo photograph

An image presented in the form of a *photograph* using data obtained by non-photographic means, for example, radio or X-ray images. Such images are often easier for the human eye/brain to interpret than contour maps, etc.

PSF

See *point spread function.*

Publications of the Astronomical Society of the Pacific

A journal published by the *Astronomical Society of the Pacific* for its members.

pulse counting

See *photon counting.*

Purkinje effect

A physiological phenomenon of the eye. The black and white receptors in the retina, the rods, peak in their sensitivity at about 510 nm, while the overall sensitivity of the color receptors, the cones, peaks at about 550 nm. Since the rods are more sensitive than the cones, faint bluish objects will be seen as brighter than they should be in comparison with bright reddish objects. The effect can cause problems when making eye estimates of stellar magnitudes.

Purple mountain observatory

An optical and millimeter wave observatory near Nanjing, China.

push-broom scan

See *scanning.*

Pushchino observatory

A radio observatory, south of Moscow, with a 22m dish.

pyrheliometer

An instrument for measuring the total energy received from the Sun.

Q

quadrant

An obsolete instrument similar to the *octant*, but with a 90° sector. See also: *sextant*.

quadrature

A geometrical alignment for an object further from the Sun than the Earth, when the Sun–Earth–object angle is 90°. See also: *aspect*.

quantum

See *electro-magnetic radiation*.

quantum efficiency

A measure of the efficiency of a detector. It is the ratio of the number of photons detected to the number of photons incident on the detector. Its value can range from 90% for *CCDs* to 0.1% for *photographic emulsions*.

quantum noise

Noise arising from the quantum nature of *electromagnetic radiation*, so that for faint sources, photons arrive in irregular bursts rather than at regular intervals, even though the source may be of constant intensity.

quarter wave plate

A device used in *polarizers* to convert *circularly* or *elliptically polarized* light into *linearly polarized* light, or vice versa. It is made from a birefringent material and designed so that the ordinary and extraordinary rays have a *phase* shift of 90° (a quarter of a wavelength) when they emerge. See also: *Pockel's cell; half wave plate*.

Questar

A commercial firm manufacturing telescopes and accessories.

Questar telescopes. (Image reproduced by courtesy of Questar Corporation.)

quick look data

Data obtained directly from an instrument, especially one on board a spacecraft, without calibration, data reduction, etc., in order to check that the instrument is working or perhaps to get an early glimpse of an exciting set of observations.

RA

See *right ascension.*

rack and pinion focuser

See *focuser.*

radar

A device comprising a radio transmitter and receiver that detects and measures distances, velocities and surface structures of objects by detecting the reflections of radiation sent out to them by the transmitter. Radar transmissions can be pulsed or continuous. The latter type, however, is not able to measure the distance of the object. Radar used for astronomical purposes may be relatively conventional equipment on board planetary space probes, or Earth-based for studying meteors. Alternatively it may be a highly specialized Earth-based system for studying the planets at long distances. The latter version is often a radio telescope that has been equipped with a transmitter, and which both sends out the pulse and receives it on its return. See also: *synthetic aperture radar.*

Radcliffe observatory

An optical observatory in Pretoria, South Africa.

radial velocity

The velocity of an object along the line of sight to it from the Earth. By convention it is positive when directed away from the Earth. See also: *Doppler shift.*

radial velocity spectroscope

See *cross correlation spectroscope.*

radian

A measure of angle, symbol, rad. 2π radians are equal to a complete circle (360°) so 1 radian = 57.296°. A useful conversion factor is 1 rad $\approx$ 200,000″ (actual figure, 206,265″). See also: *small angle relationships.*

radiance

See *surface brightness.*

R

radiant

The point in the sky from which the paths of meteors in a meteor shower appear to diverge. It represents the direction in space of the relative velocity between the Earth and the cloud of particles responsible for the shower. The constellation within which the radiant is located is used to name the shower; e.g. the Leonids, the Lyrids, the Perseids, etc. See also: *convergent point.*

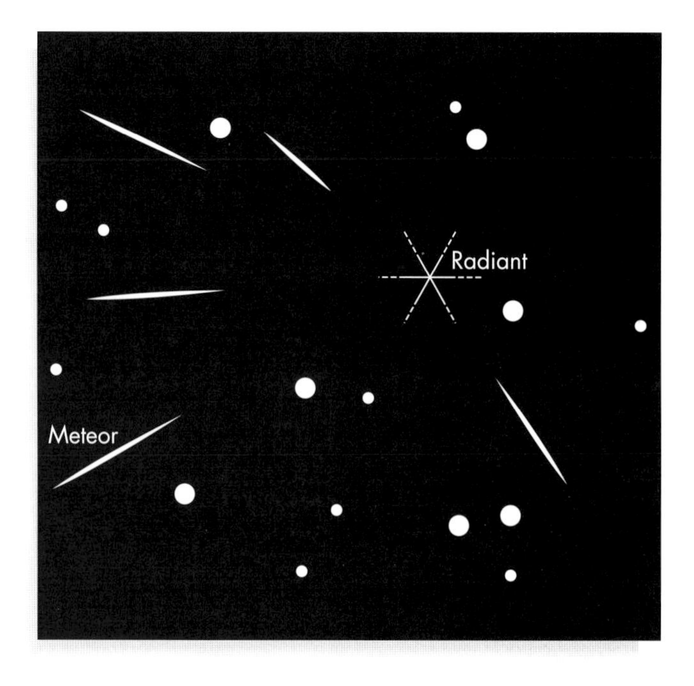

The radiant of a meteor shower.

radiation

See *electromagnetic radiation.*

radio astronomy allocations

The wavelength zones within the *radio region* that are wholly or partially reserved for astronomical use (see appendix for a full list).

radio interferometer

See *interferometer.*

radiometer

An instrument for measuring the intensity of *electromagnetic radiation* operating over any part of the spectrum. A *photometer* is a radiometer working in the optical region.

radio receiver

The part of a *radio telescope* that amplifies and converts into a useable form, the electrical signals induced into the *feed* by the incoming radio signal. Most radio receivers operate on the super heterodyne principle. This mixes the main signal with a signal from a local oscillator at a similar but different frequency. The beat frequency between the two, usually known as the intermediate frequency, is then amplified. The receiver may also filter the signal and convert it to a voltage proportional to the input power. The final output is to a chart recorder, computer, etc. See also: *antenna; Dicke radiometer.*

radio region

The longest *wavelength*, lowest *frequency* form of *electromagnetic radiation*. The wavelength is 10 mm or more and the frequency 3×10^{10} Hz or less.

radio telescope

An instrument for collecting, detecting and amplifying long wavelength *electromagnetic radiation*. The detection and amplification stages are sometimes separated out and called the *radio receiver*, while the term radio telescope is reserved for the physical structures that collect the radiation. These latter may be *parabolic dishes*, often with a secondary reflector to give a *Cassegrain design*, or they may be less conventional structures such as collections of feeds at strung along the line of focus of a parabolic cylinder, or fixed parabolas fed by an auxiliary reflector (see *Krauss radio telescope*). In almost all cases the collecting part of the radio telescope is physically large in order to concentrate the intrinsically weak signals, and to improve the angular *resolution* of the instrument. Greater resolution and sensitivity may be obtained by combining the outputs of two or more individual radio telescopes (see *interferometer*, phased *array* and *aperture synthesis*).

The 32m radio telescope dish at Cambridge that is part of the MERLIN radio interferometer. (Image reproduced by courtesy of Ian Morison, Jodrell Bank Observatory and the University of Manchester.)

radio window

The range of long wavelength *electromagnetic* waves that is able to penetrate to ground level on the Earth. It extends from about 10 mm to about 30 m. See also: *atmospheric window; optical window.*

radome

A dome protecting a *radio telescope* from the weather. It is made from material that is transparent to the operating wavelength and the radio telescope observes through the material. Unlike the *dome* for an optical telescope, it therefore does not need an opening.

The Haystack observatory with a dish enclosed in a radome in the distance. (Image reproduced by courtesy of MIT Haystack observatory.)

Ramsden disk

See *exit pupil.*

Ramsden eyepiece

A class C (see *eyepieces*) eyepiece made from two *plano-convex lenses* of equal *focal lengths* and with a separation equal to their focal length. It has a useable *field of view* of about 25°, the *eye relief* is short and there is some *chromatic aberration* and *field curvature*. The focal point of the eyepiece is in front of the *field lens*, so that it is simple to add *cross wires* to it if so desired. The standard design has the field lens at the focus of the *eye lens* so that any dust on the field lens is in focus. In practice therefore the lens separation is often made less than the focal length.

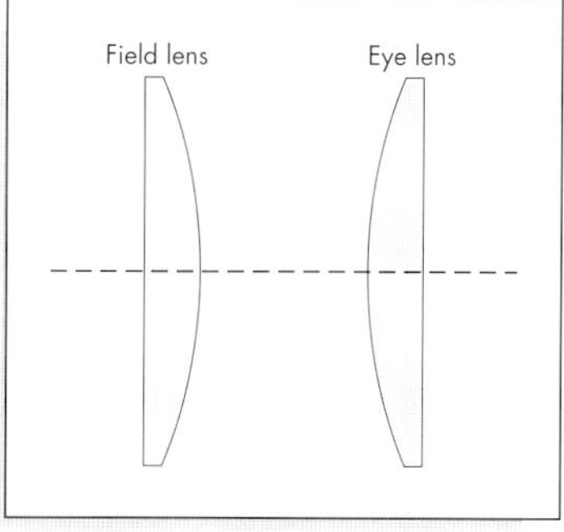

The optical arrangement of the Ramsden eyepiece.

random noise

Noise that is unpredictable with regard to its instantaneous effects and varies randomly both in magnitude and frequency. It may, however, average out to a constant level. It usually has a *Gaussian distribution*, and the *uncertainty* that it introduces to measurements is calculated through the *standard deviation* or *standard error of the mean*. See also: *systematic noise; white noise; Wiener filter.*

RAS

See *Royal Astronomical Society.*

RAS thread

See *eyepiece mounting.*

raster scan

See *drift scan.*

RATAN-600

A radio telescope in Southern Russia made of some 900 individually controled panels in the form of a ring nearly 600 m across. See also: *Zelenchukskaya observatory.*

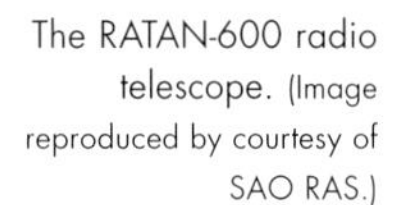

The RATAN-600 radio telescope. (Image reproduced by courtesy of SAO RAS.)

raw data

Data obtained directly from the instrument, without correction of known faults or any other form of processing.

raw image

The image straight from the instrument, without being processed in any way.

Rayleigh–Jeans approximation

A relationship that gives a good approximation to the distribution of *thermal radiation* at long wavelengths. It is simpler than the Planck equation, and so is often used at radio wavelengths. It is given by:

$$I_\lambda \approx \frac{2kc\mu^2 T}{\lambda^4},$$

where I_λ is the intensity at wavelength, λ, per unit wavelength interval, k is Boltzmann's constant ($k = 1.381 \times 10^{-23}$ J K^{-1}), c is the velocity of light in a vacuum ($c = 2.998 \times 10^8$ m s^{-1}), μ is the refractive index of the medium (normally close to 1) and T is the temperature.

Rayleigh resolution

A measure of the ability of an optical system to separate two objects. It is defined as the separation of two point sources when the center of the *Airy disk* of the image of one source is centered on the first dark fringe of the image from the second source, and vice versa. For a conventional optical telescope or a dish-type radio telescope with a circular *objective*, it is given by 1.22 λ/D *radians* (where λ is the *wavelength* and D the objective diameter). For visual telescopes, this equates to 0.12/D seconds of arc (where D is in meters). It is also known as diffraction-limited resolution. It is a useful but arbitrary definition of resolution. Practiced double star observers will often be able to better it under good condition by a factor of two or three. On the other hand, if the two objects are of very different brightnesses, then the resolution may be ten or more times worse than the Rayleigh resolution. Extended linear objects, such as the Cassini division of Saturn's rings, can often be seen more easily than the Rayleigh criterion would suggest. For all telescopes the optics must be of *diffraction limited* quality if they are to reach their Rayleigh resolution, and for optical telescopes more than about 0.3m in diameter, *seeing* within the Earth's atmosphere will normally be the limiting factor on resolution. See also: *adaptive optics; interferometer resolution; λ/4 criterion; λ/8 criterion; wave front; Fried's coherence length.*

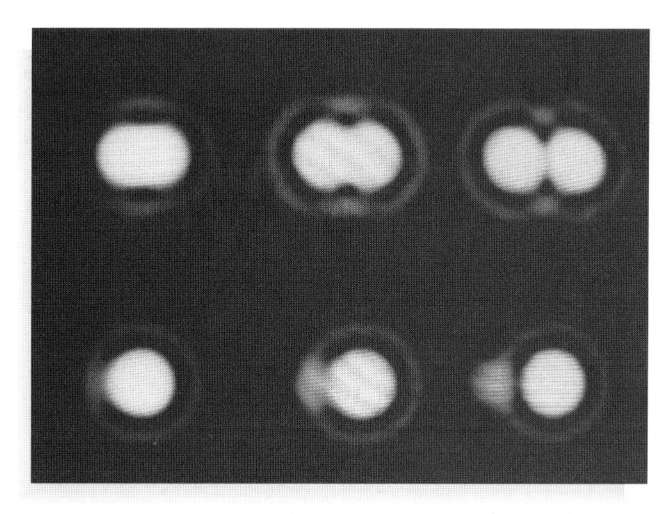

The appearance of a close double stars at separations of $\frac{2}{3}$, 1 and $\frac{3}{2}$ times the Rayleigh resolution.
Top Stars of equal brightnesses.
Bottom Stars differing by three magnitudes in brightness.

ray tracing

A computer based method for assessing designs of optical systems. The computer follows the paths of a number of parallel rays incident onto the system at some angle to the optical axis, and calculates where each ray intersects the image surface. This is repeated for a number of different angles of incidence and the results plotted out to produce a spot diagram. Ray tracing does not take account of *diffraction* effects, so a design is successful when all the spots are smaller than the *Airy disk.*

real image

An image that may be projected onto a screen. Produced, for example, when a parallel beam of light falls onto a *converging lens* or concave mirror. See also: *virtual image.*

rear illuminated CCD

See *CCD.*

reciprocity failure

The loss of speed occurring when a *photographic emulsion* is used at long exposures. The name arises because the reciprocal relationship between image intensity and required exposure length has broken down. Reciprocity failure is minimized, but not eliminated, in some emulsions produced specifically for astronomical use.

recombination

The inverse process to *ionization*, when a free electron is captured by an *ion*, with the released energy usually being emitted as *electromagnetic radiation.*

reddening

See *interstellar absorption.*

R

reddening ratio

The ratio of the U–B and B–V *color excesses*. It has a value of about 0.72 that is independent of the amount of *interstellar absorption*, and it can therefore sometimes be used to estimate the distance of an object, and its un-reddened properties.

red-dot finder

A zero power finder that provides a projected red dot (or circle, or concentric circles, etc.) onto the sky for telescope setting. See also: *Telrad finder.*

red shift

See *Doppler shift.*

reflecting telescope

See panel (page 203).

reflection grating

A *diffraction grating* where the individual apertures are long thin mirrors. Reflection gratings are produced by repeated diamond rulings across a reflective coating, usually aluminum. See also: *blazed grating; replica grating; transmission grating.*

reflective coating

See *aluminizing*, *mirror coatings*, and *silvering.*

reflector

See *reflecting telescope.*

refracting telescope

See panel (page 204).

refraction

See *atmospheric refraction* and *refractive index.*

R

refractive index

A parameter that describes the amount by which a transparent substance changes the direction of a light ray passing through it, normally symbolized by μ or n. It is defined by Snell's law:

$$\mu = \frac{\sin i}{\sin r},$$

where i is the angle of incidence (the angle between the light ray and the normal to the surface of the material) and r is the angle of refraction. It is also equal to the ratio of the velocity of light in a vacuum to the velocity of light in the material. Most optical glasses have refractive indices in the range 1.5 to 1.7.

refractor

See *refracting telescope.*

relay lens

A lens within an optical system whose sole purpose is to reverse the orientation of the image. See also: *terrestrial telescope; binoculars; Porro prism; transfer lens.*

reflecting telescope

A telescope whose primary means of collecting and concentrating radiation is by reflection from a concave mirror. The mirror is parabolic in shape in most designs, but may also be hyperbolic or spherical in others. See individual designs for further details. See also: *Cassegrain telescope; Maksutov telescope; Newtonian telescope; radio telescope; Schmidt–Cassegrain telescope; Ritchey–Chrétien telescope.*

The primary mirror of a small Newtonian reflecting telescope.

A skeleton Newtonian reflecting telescope with a second Newtonian piggy-backed as a finder/guider.

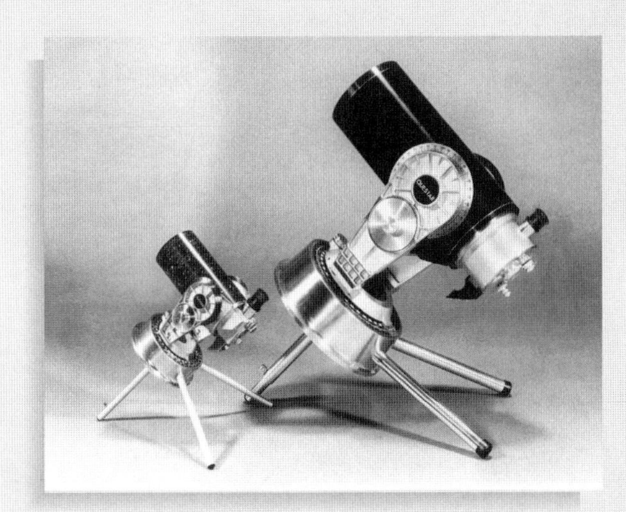

Maksutov reflecting telescopes. (Image reproduced by courtesy of Questar Corporation.)

A Schmidt-Cassegrain reflecting telescope. (Image reproduced by courtesy of Celestron International)

R

refracting telescope

A telescope whose primary means of collecting and concentrating radiation is by the use of a *lens*. The lens must be formed from at least two components in order to suppress *chromatic aberration* (see *achromatic* and *apochromatic lenses*). Refracting telescopes are limited to about 1m (40-inches) diameter because their lenses may only be supported at their edges. The astronomical refractor uses a positive (*converging*) lens or lenses as its *eyepiece*, and produces an *inverted image*. The *terrestrial* refractor additionally has a *relay lens* or *Porro prisms* to provide an *erect image*. The *Galilean telescope* uses a *negative lens* placed before the objective's focus and produces an erect image.

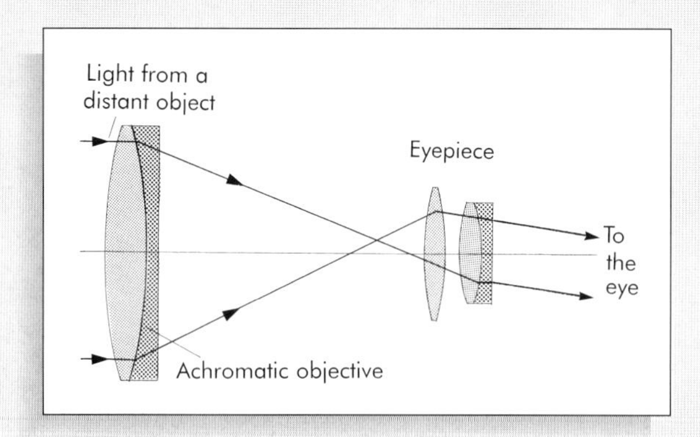

The optical system of an astronomical refractor.

A modern refracting telescope. (Image reproduced by courtesy of Meade Instrument Corp.)

The 1 m (40-inch) refractor at Yerkes.

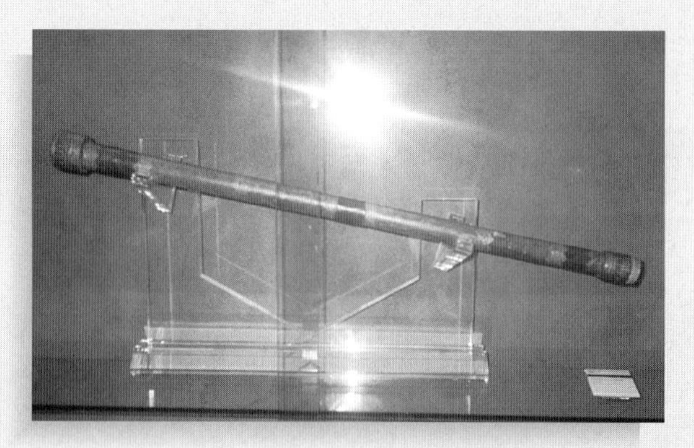

One of Galileo's refractors

R

remote observing
Observing carried out using telescopes that are controled and operated by observers some considerable distance away from the instruments. Clearly all balloon-borne and space telescopes are remotely operated, and the technique is starting to be applied to Earth-based telescopes since it can reduce costs and increase the efficient use of staff and telescope time. See also: *robotic telescope*.

remote sensing
Any technique where the information is gained purely by observational means (i.e. not by experimentation or direct intervention with the object concerned). Most astronomical work thus comes under the heading of remote sensing, but so too do radar and medical scanning techniques, etc.

replica grating
A *diffraction grating* that is manufactured as the cast of an original grating. The original grating is covered with a thin layer of a liquid plastic or epoxy resin that is stripped off and mounted onto a suitable substrate when it has set. The cast is then aluminized to give it its reflective coating. Replicas are cheaper than original gratings since many can be produced from a single original, but of slightly lower quality. See also: *reflection grating; blazed grating.*

reseau
A grid or other array of markers placed onto an imaging detector to provide points from which to measure the positions of objects within the image.

residual
The difference between a measured quantity and its predicted value. Residuals may arise through uncertainties in the data or from a real phenomenon whose effect has not been included in the calculations.

resolution
See *Rayleigh resolution*, *Dawes limit* and *spectral resolution*.

responsive quantum efficiency
A synonym for *quantum efficiency*.

responsivity
The output from a detector when the input signal has unit intensity.

retarder
A synonym for a *wave plate*.

reticle
See *graticule*.

retina
The layer at the back of the *eye* containing the light receiving cells. The retina contains two types of detectors, known from their shapes, as rod and cone cells. The cones are of three varieties, sensitive to the red, yellow and blue parts of the spectrum respectively, and enable us to see in color. The rods are sensitive to only the yellow part of the spectrum. In bright light the rods are almost depleted of their light sensitive chemical (visual purple, or rhodopsin), and we therefore see via the cones. At low light levels, the rhodopsin in the rods regenerates slowly. When the rhodopsin is fully

replaced, the rods are about one hundred times more sensitive than the cones. We then see principally via the rods in the retina. This behaviour results in two commonly experienced phenomena. The first is *dark adaptation*, the second, is *monochromatic vision*.

The rods and cones are not evenly distributed throughout the retina. The cones are concentrated in a region called the *fovea centralis*, which is the point where the image falls onto the retina when we look at something directly. The rods become more plentiful away from this region. If we look directly at a faint extended object, its image will fall onto the *fovea centralis* with its concentration of low sensitivity cones, the use of *averted vision* will then enable fainter objects to be seen by displacing the image onto the more sensitive rod cells.

There are about 10^8 rod cells and 6×10^6 cone cells in the average eye, but only about 10^6 nerves in the optic nerve. Many rods and cones therefore feed a single nerve going to the brain, resulting in phenomena such as *irradiation* and the *teardrop effect*.

retrograde motion

The "unusual" direction of motion within the solar system. For planets in the sky, it is from East to West against the background stars, for rotating or orbiting solar system objects it is clockwise as seen from above the solar system in the direction of the *north pole*. See also: *direct motion*.

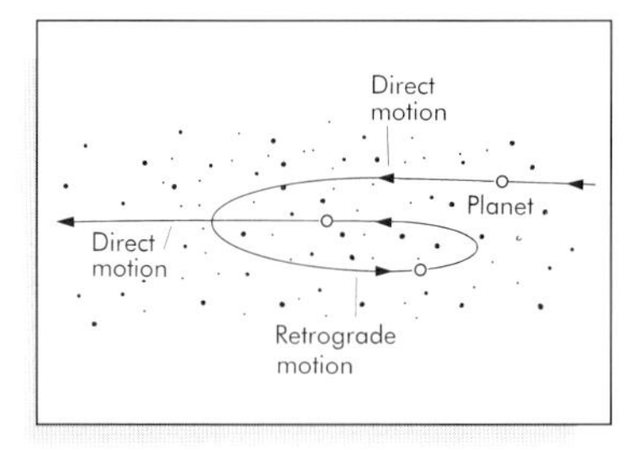

The movement of a planet across the sky (schematic) showing direct and retrograde motion.

reversal

A technique for the *photographic* production of a *positive image* directly, as for example with color slides. See also: *solarization*.

RGO

See *Royal Greenwich Observatory*.

rhodopsin

See *retina*.

R

rich field telescope

See *wide field telescope*.

right ascension

See panel (page 60).

right ascension circle

See *setting circle*.

rising

The moment of time when an object appears above the *horizon*. (Tabulated annually for the Sun and Moon in the Astronomical *Almanac*.)

Ritchey–Chrétien telescope
A variant of the *Cassegrain telescope* design that uses a hyperbolic primary mirror. The design gives a wider field of sharp focus than the Cassegrain, and is used for many modern large reflectors.

A modern 24-inch Ritchey–Chrétien telescope. (Image reproduced by courtesy of Optical Guidance Systems.)

RK eyepiece
See *Kellner eyepiece.*

RKE eyepiece
See *Kellner eyepiece.*

robotic telescope
A telescope that carries out an observing program without direct human control. Usually this will be a computer-controled instrument. See also: *remote observing.*

Rochon prism
A *polarizer* that produces two diverging beams which are linearly polarized at right angles to each other. It consists of two right-angled prisms with orthogonal optical axes, cemented together along their long sides. It is similar to the *Wollaston prism* but has the optical axis of the first prism at right angles to its orientation in the Wollaston prism. See also: *beam splitter.*

rod cell
See *retina.*

ROE
See *Royal observatory; Edinburgh.*

Ronchi grating
A grating with a spacing of a few lines per millimeter used in the *Ronchi test.*

Ronchi test
A variation of the *Foucault test* that uses a grating in place of the knife-edge. A series of lines then appears over the mirror whose shapes may be related to the shape of the mirror. See also: *Hartmann test.*

roof prism
A synonym for a *Porro prism.*

Roque de los Muchachos observatory
An optical observatory housing the *Isaac Newton group* and several other telescopes on La Palma in the Canary islands, altitude 2400 m. See also: *William Herschel telescope.*

rotational broadening

See *Doppler broadening*. See also: *line broadening*.

rotational transition

See *transition*.

rotation of the field of view

See *image rotation*.

Rowland circle

The basis for several designs of *spectroscope* that use curved *diffraction gratings*. The Rowland circle is simply the continuation of the shape of the grating, and the final spectrum follows the line of the circle when the spectroscope entrance aperture is also positioned on the circle.

Rowland ghost

See *ghost*.

Royal Astronomical Society

A UK-based society for amateur and professional astronomers. See http://www.ras.org.uk/.

Royal Greenwich observatory

The organization and research center that managed many of the UK's telescopes until 1998, when that function was passed to the *Royal observatory; Edinburgh*. It was originally based at the *Royal observatory; Greenwich*. It moved to the original site of the *Isaac Newton telescope* at Herstmonceux in Sussex, and then to Cambridge, next to the Institute of Astronomy, before being closed in 1998.

Royal observatory, Edinburgh

An optical observatory sited at Edinburgh, United Kingdom. Also the organization that manages the UK's major telescope sites and, through the Astronomy Technology Centre, the UK's national center for the design and production of astronomical equipment.

Royal observatory, Greenwich

An optical observatory, now a museum, sited at Greenwich, London. It was founded in 1675 and housed the Airy transit circle. The center line of the latter defines the Greenwich meridian from which all longitudes on the Earth are measured.

Rozhen observatory

An optical observatory with a 2m *Ritchey-Chrétien* and other telescopes sited at an altitude of 1800 m in Bulgaria.

run-off roof

A roof for an observatory building. The roof is mounted on wheels so that it may be moved along tracks (run-off) in order to open the building for observing. See also: *dome*.

S

Sacramento peak observatory
See *National solar observatory.*

SAGE neutrino detector
A *neutrino detector*, based upon detecting the conversion by a neutrino of gallium-71 to radioactive germanium-71. It uses 60 tonnes of liquid gallium and is buried beneath the Caucasus mountains.

Saha equation
An equation that determines how the relative number densities of atoms in various levels of *ionization* change with temperature. It takes the form:

$$\frac{N_{I+1}}{N_I} = \frac{2U_{I+1}}{U_I}\left(\frac{2\pi m_e kT}{h^2}\right)^{3/2} e^{-\chi_I/kT},$$

where the subscript I or I+1 denotes the level of ionization, N is the number density, U is an atomic constant called the partition function, m_e is the electron mass (9.11×10^{-31} kg), k is Boltzmann's constant (1.38×10^{-23} J K^{-1}), T is the temperature, h is Planck's constant (6.63×10^{-34} J s) and χ_I is the energy required to ionize the Ith ion to the $(I+1)^{th}$ ion. See also: *ion.*

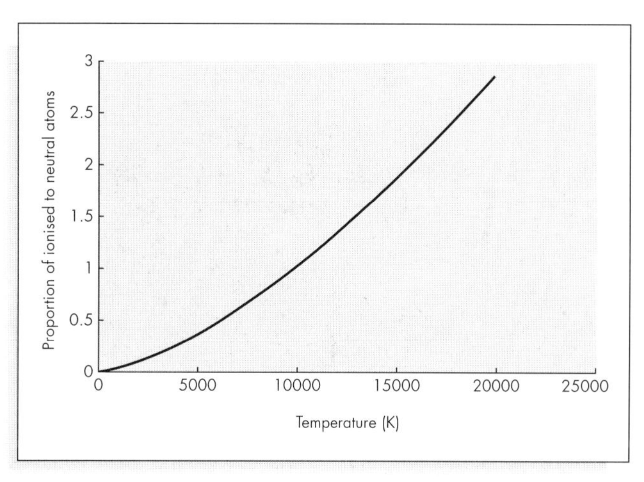

The variation in the ratio of ionized to neutral atoms as predicted by the Saha equation (example only, not for a specific atom).

SALT
See *South African large telescope.*

sampling theorem
The theorem that, when digitising a signal, the highest frequency that will be correctly measured is half the sampling frequency. See also: *Nyquist frequency; aliasing.*

Santa Barbera Instrument Group
A commercial firm manufacturing *CCD* cameras, spectroscopes, etc. for astronomical use.

SAO star catalog
A catalog of the positions and proper motions of over a quarter of a million stars, produced by the *Smithsonian Astrophysical observatory* in 1966.

SAR
See *synthetic aperture radar.*

saros
The interval of 18 years and 10 or 11 days after which a solar or lunar eclipse recurs under similar circumstances to the previous occasion. The interval exceeds an exact number of days by 8 hours, and so each successive eclipse is moved 120° around the Earth. After three saros intervals, the eclipse recurs near the same geographical point on the Earth.

satellite

1 A small naturally-occurring body orbiting a planet or asteroid. (The positions of many satellites of the planets are tabulated annually in the Astronomical *Almanac.*)
2 Any spacecraft whether in orbit around a planet or journeying through the solar system.

saturation

The maximum response of a *detector*. The output becomes constant as the input increases. Often the detector response will become non-linear before saturation is reached, so that the useful range of the detector ceases at some point before it saturates. See also: *dynamic range.*

SBIG

See *Santa Barbera Instrument Group.*

S Cam

See *superconducting tunnel junction detector.*

scanning

A technique for obtaining an image of an extended source using a point detector. The image is moved over the detector, or vice versa, in a regular pattern, the output from the detector is recorded, and the image synthesized from the recording. The commonly used scanning patterns are the *drift scan* or a square spiral out from the center point. Push broom scanning often produces images from spacecraft. This has a linear array of detectors at right angles to the trajectory of the spacecraft, and these record continuously to build up the image of the strip of land below the spacecraft. See also: *coded array mask; Hadamard mask imaging.*

Schiefspiegler telescope

A *reflecting telescope* in which the light path is not obstructed by the secondary mirror. The spheroidal *primary mirror* is tilted so that the secondary mirror is to one side of the light path. In the Kutter design, the *coma* and *astigmatism* introduced into the image by tilting the primary mirror are partly compensated by tilting the concave spheroidal secondary mirror, and further reduced by using long (f20) focal ratios. In the Yolo design, the mirrors are stressed to deform their shapes in such a way that the aberrations are reduced or eliminated. There are many other variations on the design. See also: *Herschelian telescope; tri- Schiefspiegler telescope.*

A Schiefspiegler telescope. (Image reproduced by courtesy of Prof. Joseph Sylvan.)

Schmidt camera

A *camera* used for direct astronomical imaging and also within many *spectroscopes.* It was designed by Bernhard Schmidt in 1930, it has a short focal ration (f2 to f1) and has, for an astronomical instrument, a wide field of sharp focus of 6° or more across. Using a spherical primary mirror eliminates most of the aberrations. The spherical aberration intrinsic to this shape is then corrected through the use of a thin *correcting lens* of complex shape placed at the radius of curvature of the primary mirror. The lens is too thin to introduce a significant *chromatic aberration.* There is one remaining aberration; *field curvature*, and this is eliminated by curving the photographic plate over a moulded support to the shape of the focal surface. Since the focus is inaccessible, the design cannot be used visually.

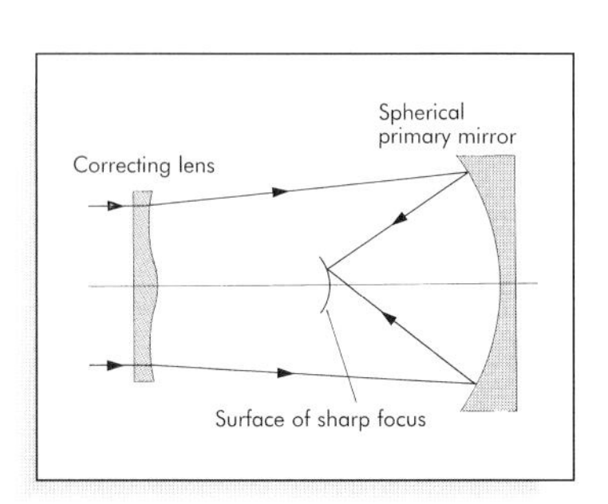

Light paths within a Schmidt camera.

Schmidt–Cassegrain telescope

A combination of the *Schmidt camera* and *Cassegrain telescope* designs that can be used visually, often abbreviated to SCT. The *correcting lens* is placed near the focus of the *primary mirror* and a convex spherical secondary mirror is mounted onto the correcting lens. The reflected beam exits through a hole in the primary mirror to a focus at the back of the telescope as in the Cassegrain design. The SCT is a very popular design for commercially produced small telescopes (see *Celestron* and *Meade*) since it produces a very compact instrument. The optical performance, however, is not as good as the Schmidt camera. See also: *catadioptric telescope; Maksutov telescope.*

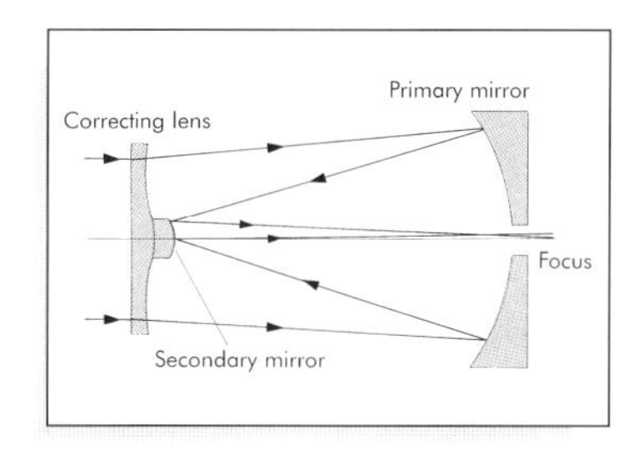

Light paths within a Schmidt–Cassegrain telescope.

Science

A USA-based science and astronomy research journal.

Scientific American

A monthly USA-based popular science magazine.

scintillation

1 The shimmering of astronomical images, especially those of stars, arising from variations in density within the Earth's atmosphere. The term scintillation may be used for the whole phenomenon, or it may refer to the high altitude component. In the latter case the low altitude component is called the *seeing*. Both effects cause the image to change its size, shape, position and brightness on a time scale of a milli second to a second. It is the primary cause of large telescopes not reaching their *Rayleigh resolutions*. See also: *Antoniadi scale; adaptive optics; Fried's coherence length.*
2 The flashes of light produced by the ionising effect of *X-rays*, *gamma rays* and *cosmic rays* as they pass through some materials. See also: *scintillation detector.*

scintillation detector

A detector for X-rays and gamma rays. Normally the detector uses sodium iodide or caesium iodide that emit visible radiation (*scintillate*) in response to the ionization produced by the passage of the high-energy photon through the material. The visible photons are then detected by a *photomultiplier*. Two superimposed layers of sodium iodide and caesium iodide enable the direction of travel of the radiation to be determined, since the pulse decay times differ between the two substances. This latter type of detector is known as a phoswich.

S

scotch mounting

See *barn door mounting.*

scotopic vision

Vision under low levels illumination, arising from the rod cells in the *retina*. See also: *photopic vision.*

SCT

See *Schmidt–Cassegrain telescope.*

second

Apart from being the familiar unit of time, it is also used as a measure of angle. 1s = 15″. This arises because objects in the sky move through 15″ in one second of *sidereal time*. See also: *hour; minute.*

secondary cosmic ray

See *cosmic ray.*

secondary mirror

A subsidiary and usually relatively small mirror in a telescope's optical train. Its function may be simply to bring the focus to a convenient position, as in the *Newtonian telescope* and the *Coudé* and *Nasmyth* designs. In which case it is a flat mirror. Alternatively it may play a part in producing the image, as in the *Cassegrain* and related designs where the secondary mirror not only reflects the light to a convenient position at the back of the primary mirror, but also increases the *focal ratio* and *focal length* of the telescope over that of the *primary mirror.* In this case the secondary mirror has a curved surface.

secondary spectrum

The remaining *chromatic aberration* for an *achromatic lens.* The achromat reduces, but cannot eliminate completely, the chromatic aberration of a simple lens, and so the secondary spectrum shows up as faint colored fringes around bright stars.

secondary star

See *double star.*

second of arc

See *arc second.*

seeing

The shimmering of astronomical images, especially those of stars, arising from variations in density at a low level within the Earth's atmosphere. The term *scintillation* may be used for the high altitude component. See also: *Antoniadi scale; adaptive optics; Fried's coherence length.*

segmented mirror

A mirror, especially the *primary mirror* of a large telescope that is made up from several smaller mirrors. The subsidiary mirrors must be kept in their correct mutual alignments, and this is done using *active supports.* The final mirror is lighter and cheaper than a monolithic mirror of the same size. See also: *honeycomb mirror; thin mirror; spun-cast mirror; Keck telescope; Hobby–Eberly telescope.*

Seidel aberration

See *aberration* (optical).

semi major axis

Half the longest diameter of an ellipse.

semi minor axis

Half the shortest diameter of an ellipse.

sensitivity

The weakest signal that can be pick-up by a detector. The noise level within the detecting system often determines its value since a *signal-to-noise level* of 1 or more is needed for reliable detection.

separation

The angular distance between two objects, such as the components of a *double star*, or a planet and one of its satellites. See also: *position angle.*

Serrurier truss

A widely used structure for large telescope "tubes". It has a central box section with each support running diagonally from one corner of the box to meet the corresponding support from the next corner of the box at the edge of the *mirror cell*. There are thus eight supporting arms for the mirror cell arranged in four triangles. The design is used because it is easy to calculate the flexure under changing gravitational loadings as the telescope points to different parts of the sky. The Serrurier truss supports for the *primary* and *secondary mirrors* may thus be designed to flex by equal amounts, keeping the mirrors in their correct alignment. Modern computer-based design packages mean that more complex and better support systems can now be used.

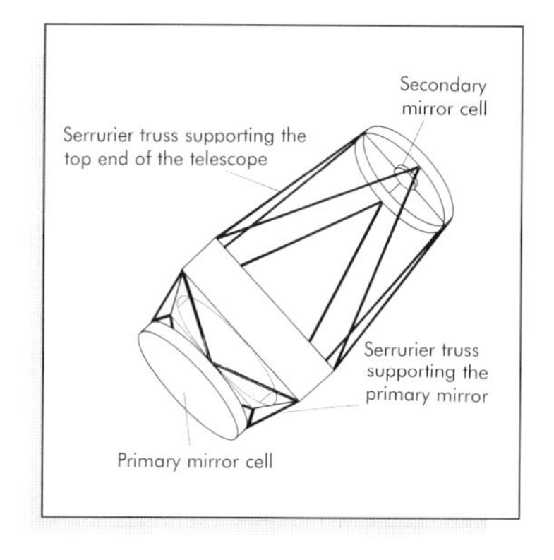

A telescope tube design based upon Serrurier trusses.

setting

1 The initial stage of *finding*, when the telescope is pointed approximately in the correct direction. See also: *slewing*.
2 The moment of time when an object passes below the *horizon*. (Tabulated annually for the Sun and Moon in the Astronomical *Almanac*.)

setting circle

A graduated angular scale on a telescope axis from which the *right ascension*, *declination*, *hour angle*, *altitude*, *azimuth*, etc. on which the telescope is set may be determined.

setting motion

A means of adjusting or moving a telescope's position by a large amount in order to move from one part of the sky to another. See also: *finding*.

sextant

An instrument for measuring the *altitude* of objects in the sky, especially the Sun for navigational purposes. It is the modern version of the *octant* and *quadrant*. It has a 60° sector that is graduated from 0° to 120°. The horizon is viewed directly, while Sun is seen via two mirrors and through appropriate filters. The top mirror is mounted onto a pivoted arm. The horizon is sighted through the small telescope incorporated into the device, and the pivoted arm moved until the solar image is aligned with the horizon. The altitude of the Sun may then be read from the scale. (CAUTION – always take appropriate precautions when observing the Sun or damage to the eye and/or telescope can result).

shade number

See *optical density*.

shade tube

See *baffle*.

Shanghai observatory

An optical and radio observatory near Shanghai, China, with a 25m dish and several other instruments.

sharpening
An *image-processing filter* that emphasizes small details.

shearing interferometer
A detector for the deviations arising from atmospheric turbulence in the *wave front* of light from a guide star, used within an *adaptive optics* system. It uses a *beam splitter* and recombines the two beams with a slight displacement. The resulting *interference* fringes then provide the details of the wave front distortions. See also: *Hartmann sensor; isoplanatic patch.*

shielding
The protection of a detector from unwanted signals, especially background noise. A passive shield simply absorbs or reflects the unwanted signals away from the detector. An active shield does not stop the unwanted signals, but identifies those responses from the detector that correspond to noise so that they may be discarded. For example a *scintillation detector* can be made directionally sensitive by surrounding it with other scintillation detectors except in the desired direction and eliminating responses from the main detector that occur simultaneously with a response from one of the surrounding detectors.

shot noise
See *noise.*

shutter
See *dome.*

side lobe
A sensitivity of a *radio antenna* or *telescope* to signals at a significant angle away from the *main lobe.* Shown on a polar diagram as a small lobe or lobes next to the main lobe.

sidereal clock
A clock adjusted to go through 24 hours in a *sidereal day.* It thus gains 3 m 55.9 s a day compared with a normal clock.

sidereal day
The interval between successive passages of the *First Point of Aries* across the *prime meridian.* It is 23 h 56 m 4.1 s long, and the difference from the *solar day* arises because of the Earth's orbital motion which results in the Sun moving with respect to the background stars. Because of *precession* it is about 0.01 seconds shorter than the Earth's actual rotation period.

sidereal period
The actual orbital period of a planet, satellite, etc. with respect to the background stars. See also: *synodic period.*

sidereal rotation period
The actual rotational period of a body with respect to the background stars. See also: *synodic rotation period.*

sidereal time

The *hour angle* of the *First Point of Aries* (tabulated annually for Greenwich in the Astronomical *Almanac*). Effectively it is time given by the motion of the stars rather than by the Sun. Since the *sidereal day* is 3 minutes 55.9 seconds shorter than the solar day, a sidereal clock will gain by that amount every day when compared with a normal clock. The sidereal time and *civil time* (ignoring summer time adjustments, etc.) agree with each other at the *autumnal equinox* (not the *spring equinox* because of the 12 hours added to the hour angle of the *mean Sun* to give civil time). Thereafter the sidereal clock gains and is 6 hours ahead at the *winter solstice*, 12 hours ahead at the spring equinox, 18 hours ahead at the *summer solstice* and back in agreement with civil time again at the next autumnal equinox, having gained a full 24 hours. For a specific observer, sidereal time, ST, is related to the right ascension, RA, and hour angle, HA, of an object by $ST = RA + HA$. See also: *right ascension; apparent sidereal time; mean sidereal time; equation of the equinoxes.*

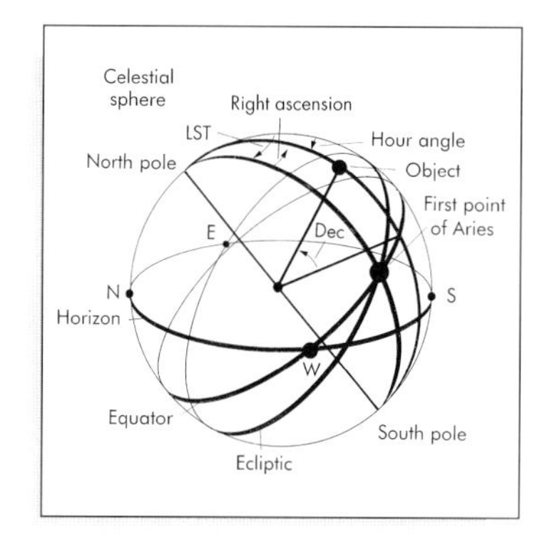

The definition of local sidereal time (LST).

sidereal year

The *sidereal orbital period* of the Earth, equal to 365.25636 days. It differs from the *tropical year* because of *precession*.

sideriostat

See *coelostat*. See also: *heliostat.*

Siding Springs observatory

See *Mount Stromlo observatory.*

signal/noise ratio

A measure of the certainty with a signal has been detected, often written as s/n. The detector output normally has to be at least double the *noise* level (s/n = 1) before the signal can be regarded as detected. If accurate measurements are to be made on the signal then s/n needs to be much larger than 1. Ratios of 100 to 500 are not uncommon for precise work. See also: *sensitivity; uncertainties.*

significance level

A statistical quantity used in tests such as *chi-squared*, and *Student-t* to quantify the results of that test. It has generally expressed as the probability that a quantity has NOT changed, or data sets are NOT correlated, etc. Thus if two sets of measurements of the brightness of a star have a 1% significance level from the Student-t test, it implies that there is a 99% probability that the star's brightness HAS changed.

significant figures

The number of figures in a measurement that are justified by the level of *uncertainty* in that measurement.

silvering

The application of a thin reflecting coat of silver to the surface of an optical component to produce a mirror. A chemical process may do this. Silver has a very high reflectivity when first applied, but tarnishes quickly. *Aluminizing* is therefore now the main method of producing mirrors. See also: *speculum metal; mirror coating.*

SIMBAD

A database containing information on over 2.8 million objects. The name derives from; Set of Identifications, Measurements and Bibliography for Astronomical Data. It is held at the *Centre de Données Astronomiques de Strasbourg* and may be accessed at http://simbad.u-strasbg.fr/Simbad. See also: *Astronomical Data Center; National Space Science Data Center; digital sky survey.*

single lens reflex camera

A photographic *camera* in which the viewfinder uses the camera lens to obtain its image, often abbreviated to SLR camera. This type of camera has the advantage of showing in advance the exact image that will be obtained. If a camera is to be used on a telescope then an SLR camera is almost essential so that the image may be focused and positioned correctly. However, the SLR camera requires the use of a mirror to divert the light into the viewfinder. That mirror is flipped out the way when the image is obtained. The movement of the mirror can often be sufficient to cause shake if the camera is attached to a telescope, and so ruin the image. A few SLR camera designs allow the mirror to be locked out of the way before the exposure is taken, but for much astronomical work a *top hat exposure* will be needed.

singlet

A single *spectrum line*. See also: *doublet; triplet.*

Sirius Instruments

A commercial firm manufacturing *CCD* cameras for astronomical use.

Sky and Telescope

A monthly USA-based popular astronomy magazine. See http://www.skypub.com.

The popular astronomy magazine *Sky and Telescope*.

S

sky background

The brightness of a portion of the sky that does not contain any detectable objects. The emission is due to scattered solar and artificial light, the integrated effect of objects too faint to be detected individually, direct emissions from the atmosphere, and from material in interplanetary and interstellar space. The *surface brightness* of the sky background can be as low as 23 magnitudes per square second of arc for an excellent site, but is often many times worse than that. The sky background is the limiting factor in the detection of an object. Generally an extended object will need a surface brightness better than that of the background sky by at least 0.8 magnitudes per square second of arc to be detected at all, and much better than that for any details to be seen. See also: *noise.*

slewing

A rapid movement of a telescope over an appreciable angle. See also: *setting; slow motion.*

slit-less spectroscope

A *spectroscope* without a slit as its entrance aperture. When such a spectroscope is used to observe a hot gaseous object, such as an H II region or the solar chromosphere, the spectrum consists of a series of images of the object in each of its *emission lines*.

Sloan digital sky survey

A sky survey currently underway, based at *Apache point observatory*. It is intended to cover 100 million objects over about a quarter of the sky. See http://www.sdss.org/. See also: *digital sky survey*.

slot

See *dome*.

slow motion

A means of adjusting or moving a telescope's position by a small amount to center an object in the *field of view* and thereafter to *guide* it whilst an image is being obtained. See also: *tracking*.

SLR camera

See *single lens reflex camera*.

small angle relationships

When an angle involved in a formula is small (<5°), and measured in *radians*, the following approximate relationships (which become more accurate as the angle decreases in size) are frequently useful: $\sin\theta \approx \tan\theta \approx \theta$ and $\cos\theta \approx 1$.

smart drive

A synonym for *periodic error correction*.

Smithsonian astrophysical observatory

An organization belonging to the Smithsonian institution that runs the *Fred Lawrence Whipple observatory*, and on behalf of the IAU runs the *Minor Planet Center* and the *Central Bureau for Astronomical Telegrams*.

S

smoothing

An *image-processing filter* that reduces random *noise* by averaging the intensities of *pixels* within a certain radius.

s/n

See *signal to noise ratio*.

Snell's law

See *refractive index*.

SNO neutrino detector.

A *neutrino detector*, based upon detecting the Čerenkov radiation from high-energy electrons produced in 1000 tonnes of heavy water in an underground tank at Sudbury in Canada.

SNU
See *solar neutrino unit.*

SOAR telescope
See *southern astrophysical research telescope.*

sodium iodide detector
See *scintillation detector.*

soft X-ray region
The long wavelength (low energy) end of the *X-ray region*, extending approximately over the wavelengths 1 nm to 0.4 nm.

solar blind photomultiplier
A *photomultiplier* that is sensitive only to *ultraviolet radiation.*

solar day
The interval between successive passages of the Sun or the *mean Sun* across the *prime meridian*. The former gives the apparent solar day, whose length varies by up to ± 20 seconds because of the variable speed of the Sun across the sky. The latter gives the mean solar day with a fixed length of 24 hours and upon which our *civil timekeeping* is based.

solar diagonal
See *Herschel wedge.*

solar eclipse
The passage of the Moon in front of the Sun, obscuring all or part of its surface. During a partial *eclipse* only a part of the solar photosphere is covered, while the Moon hides the whole of it during a total eclipse, allowing the chromosphere and corona to be seen. If the Moon is towards apogee when the eclipse occurs, then its disk may have a smaller angular size than that of the Sun, and an annular eclipse occurs, with a bright ring of the photosphere still visible even when the Moon is centered on the Sun. Between 2 and 5 solar eclipses occur each year.

CAUTION – always take appropriate precautions when observing the Sun or damage to the eye and/or telescope can result.

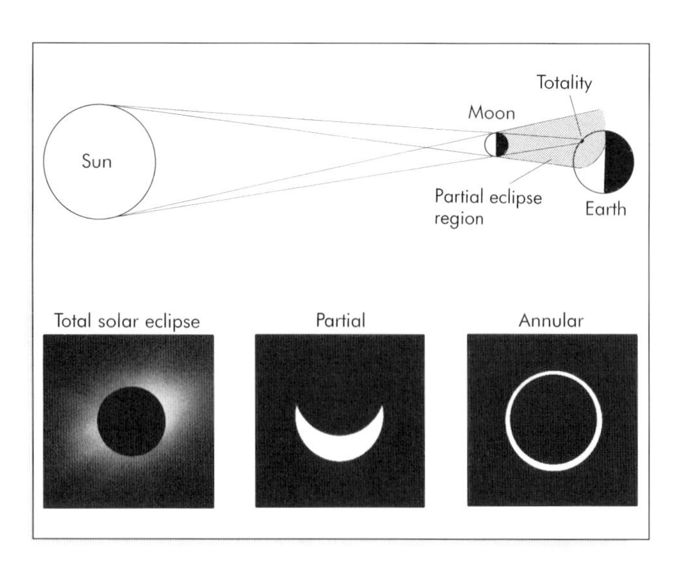

Solar eclipses.

solar eclipse viewer
Any device designed for watching the progress of a solar eclipse. In particular the term is often used for cardboard spectacles that have plastic film *solar filters* in place of the normal lenses. Such spectacles are often commercially-produced and sold cheaply around the times of solar eclipses.

CAUTION – always take appropriate precautions when observing the Sun or damage to the eye and/or telescope can result.

solar filter

A *filter* designed to permit safe observation of the Sun. They are also often called full aperture filters because they cover the whole aperture of the telescope. On large telescopes, the telescope must be *stopped down* to the size of the filter with an opaque screen. The filters reduce the intensity of the solar light to less than 0.003% of its normal intensity across the ultraviolet, visible and infrared regions. This corresponds to an *optical density* of at least 4.5 or to a *shade number* of 11 or more. They are a coating of a metal such as aluminum or stainless steel on a very thin plastic film or an optical quality glass sheet.

Aluminized-mylar solar filters.

CAUTION – always take appropriate precautions when observing the Sun or damage to the eye and/or telescope can result. Always follow the filter manufacturer's instructions, and never use a filter at the eyepiece end of the telescope. Always use a filter that has been manufactured for solar observing. The following substitutes are sometimes suggested, especially when there is a solar eclipse, but they are NOT safe: smoked glass; exposed black and white film; CDs; space blankets; aluminized helium balloons; aluminized potato crisp packets; floppy disks; crossed Polaroid filters; smoked plastic; sun glasses; mirrors; X-ray photographs. See also: *H-α filter; solar eclipse viewer.*

solarization

An effect in *photography* where at very high levels of illumination, the *characteristic curve* starts to slope downwards. The effect may be used to produce *positive images* directly. See also: *reversal.*

solar neutrino unit

A unit for measuring the flux of neutrinos from the Sun. It is the flux of neutrinos that would produce one interaction per second amongst 10^{36} chlorine-37 atoms (since the Homestake *neutrino detector* uses chlorine).

solar observing

The observation of the Sun and its features by one of the safe procedures. The main such practices are the use of a *solar filter* or *eyepiece projection* (provided that your telescope is suitable – check with the manufacturer first). Instruments specifically designed for solar work include *H-α filters*, *prominence spectroscopes*, *spectrohelioscopes* and *coronagraphs*. Some professional observatories have telescopes specifically designed for solar work (see *Big Bear observatory* and *McMath-Pierce solar telescope*).

CAUTION – always take appropriate precautions when observing the Sun or damage to the eye and/or telescope can result. See also: *tower solar telescope; solar telescope.*

solar parallax

See *diurnal parallax.*

solar telescope

A telescope optimized for solar observation. The requirements for such a telescope include having as few optical components as possible to reduce scattered light, a sealed tube that is helium-filled or evacuated to minimize convection currents, and siting at a high altitude. Most solar telescopes have objectives with long focal lengths and high focal ratios so that a large image of the sun is obtained without the use of subsidiary optics. For teaching purposes and group observation, small solar telescopes with folded light paths are produced commercially.

CAUTION – always take appropriate precautions when observing the Sun or damage to the eye and/or telescope can result. See also: *Big Bear observatory; McMath–Pierce solar telescope; tower telescope.*

An ingenious solar telescope with a folded light path for teaching and demonstration purposes. (Image reproduced by courtesy of Learning Technologies Inc.)

solar time

The time given by the real Sun and by a *sundial*. It is defined as the *hour angle* of the Sun plus 12 hours. Its difference from *mean solar time* is given by the *equation of time.*

solid angle

The two-dimensional equivalent of an *angle*. A "normal" angle is defined (in units of *radians*) as the length of the sector of a circle enclosed by the angle divided by the radius of the circle. A solid angle is similarly defined as the area of the portion of a sphere enclosed by the solid angle divided by the radius of the sphere squared. Its units are steradians, with a complete sphere having a solid angle of 4π steradians. A square in the sky whose sides are each 1° long has a solid angle of 0.000305 steradians.

solid Schmidt camera

A variant of the *Schmidt camera* that is formed from a solid piece of glass. It can reach focal ratios as low as f0.4, and is used in some *spectroscopes.*

solid state detector

A detector whose operating principle depends upon the properties of solid-state devices. The *CCD* is a solid-state detector, but the term is more usually applied to detectors for X-rays and gamma rays. For example, *pair production* in germanium by MeV gamma rays.

solstice

One of two points in the sky where the ecliptic is at its highest or lowest *declination* (± 23° 27') and hence at its greatest distance from the *celestial equator*. The Sun passes through these points on or about 21 June (summer solstice) and 21 December (winter solstice) each year. See also: *First Point of Aries; autumnal equinox.*

South African astronomical observatory

An optical observatory at Sutherland in South Africa, altitude 1800 m. It will house the *South African large telescope* and currently has several smaller instruments.

South African large telescope

An 11m optical telescope with a spherical mirror formed of 91 segments held in position by *active supports*. It will operate in a similar manner to the *Hobby–Eberly* telescope. It is sited at Sutherland on the Great Karoo plateau in South Africa, and is currently under construction for completion in 2005. See homepage at http://www.saao.ac.za/salt/.

southern astrophysical research telescope

A 4.2m optical and infrared *Ritchey–Chrétien telescope* under construction on *Cerro Pachón.*

southern lights

See *aurora.*

Soviet 6-meter telescope

See *Bolshoi telescop azimutalnyi.*

Space Telescope Science Institute

A *NASA* organization that runs the research program for the *Hubble space telescope,* based in Baltimore, Maryland.

spacecraft trail

See *trails.*

spark detector

A detector for gamma rays and cosmic rays. Its operating principle is closely related to that of the Geiger counter. It has a number of parallel plates that are charged to successively higher voltages. The passage of an ionising photon or particle leads to an avalanche of electrons along the path that may be seen as a series of sparks between the plates.

spatial filter

1 An *image processing* technique. The *filter* changes the intensity of a given *pixel* by combining it with the intensities of surrounding pixels. The action of the filter may be represented as 3×3, 5×5, 7×7, etc. matrices. The elements of the matrix are the weightings given to each pixel intensity when they are combined and used to replace the original intensity of the central pixel. Filters can act to *smooth* or *sharpen* an image, or enhance edges, etc.
2 A mask in an optical system (see *cross correlation spectroscope* and *Hadamard mask imaging*).
3 A mask in an optical system using coherent (laser) radiation that filters out *spatial frequencies* in the original image.
4 A mathematical technique that filters out spatial frequencies in the *Fourier transform* of the original image.

spatial frequency

The frequency with which a cyclic variation in an image repeats itself, per unit distance across the image. For example, in an image produced by combining several linear scans, the number of such scans per meter on the image (or centimeter, millimeter, etc.) would be its spatial frequency.

speckle interferometry

A technique that allows large Earth-based telescopes to operate at close to their theoretical *resolution.* Many images of the object are obtained on a large *plate scale* and with an exposure short enough (0.001 s) to freeze the atmospheric *scintillation.* Each image is then composed of many speckles that are *interference* patterns resulting from light passing through different cells within the atmosphere. The images are added together and optically reconstituted to produce the high-resolution image. The technique is most used for studying close double stars and for the direct imaging of nearby red giants.

spectral class

See *spectral type*.

spectral index

A measure of the way that the intensity of a radio source varies with frequency. For most astronomical sources the intensity increases or decreases exponentially, and the spectral index is the exponent of that variation. For *synchrotron radiation* the spectral index is negative so that the intensity increases as the frequency decreases and it has a value between 0 and −1.5. The index for *thermal radiation* usually lies between 0 and 2.

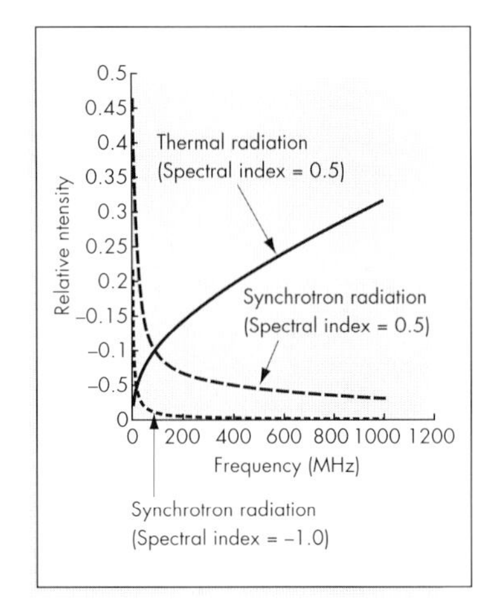

Spectral index.

spectral order

For a spectrum produced by a *diffraction grating*, it is the number of whole wavelengths difference between the path lengths from successive apertures of the grating to the spectrum. Conventional *spectroscopes* usually use first or second order spectra (1 or 2 wavelengths difference). But *echelle grating*-based spectroscopes can use orders of a 100 or more.

spectral resolution

The ability of a *spectroscope* to separate two close wavelengths. It is usually defined as:

$$R = \frac{\lambda}{\Delta\lambda},$$

where λ is the wavelength and $\Delta\lambda$ the minimum wavelength separation that can be distinguished. A *prism*-based spectroscope might have a resolution in the region of 1500. The resolution of a *diffraction grating* is given by mN, where m is the *spectral order* and N, the number of rulings in the grating. Spectral resolutions for diffraction grating-based spectroscopes thus range from 1000 or 2000 for conventional gratings to 100,000 for *echelle gratings*.

S

spectral type

A temperature classification of the stars based upon the appearance of their *spectra* in the blue-green region. The scheme originated in Annie Jump Cannon's work at Harvard in the early 1900s and was called the Harvard system. It was extended in 1943 by William Morgan and Philip Keenan who introduced the *Luminosity Class*, so that it is now generally called the MK system (sometimes known as the MKK system after Edith Kellman, who also worked on it).

There are seven major groups in the core of the classification system, labelled with the letters O B A F G K M in order of decreasing temperature. The mnemonic "Oh Be A Fine Girl/Guy Kiss Me" may be found useful for remembering the sequence. Each of these major groups is subdivided into ten by the addition of an Arabic numeral between 0 and 9.

The hottest stars so far discovered that are classified in this system are O4 stars at about 40,000 K, and the coolest, M8 stars at about 2,500 K. The Sun, with a surface temperature of about 5,700 K is classified as a G2 star.

Recently the original decimal subdivision of the core classes has been adapted to provide a smoother variation with temperature by adding and deleting some of the classes. The full range of spectral types in current use is thus:

Core	Subdivisions
O	4, 5, 6, 7, 8, 9, 9.5
B	0, 0.5, 1, 2, 3, 5, 7, 8, 9.5
A	0, 2, 3, 5, 7
F	0, 2, 3, 5, 7, 8, 9
G	0, 2, 5, 8
K	0, 2, 3, 4, 5
M	0, 1, 2, 3, 4, 7, 8.

See also: *luminosity class.*

Spectrasource Instruments

A commercial firm manufacturing *CCD* cameras for astronomical use.

spectrograph

See *spectroscope.*

spectroheliogram

A monochromatic image of the Sun obtained by a *spectrohelioscope*, and usually at the wavelength of a strong absorption line such as calcium H or K (396.8 nm and 393.3 nm) or H-α (656.3 nm).

CAUTION – always take appropriate precautions when observing the Sun or damage to the eye and/or telescope can result.

spectrohelioscope

An instrument for producing a monochromatic image of the Sun. It is a *spectroscope* with a second slit placed at the spectrum to isolate the desired wavelength. The entrance slit is oscillated so that it scans over the solar disk. The second slit is oscillated in phase with the entrance slit so that it remains aligned on the desired wavelength. If the frequency of the oscillations is 10 Hz or higher, then an image of the Sun may be seen directly. Alternatively the image may be photographed or recorded with a CCD detector.

CAUTION –always take appropriate precautions when observing the Sun or damage to the eye and/or telescope can result. See also: *H-α filter; prominence spectroscope.*

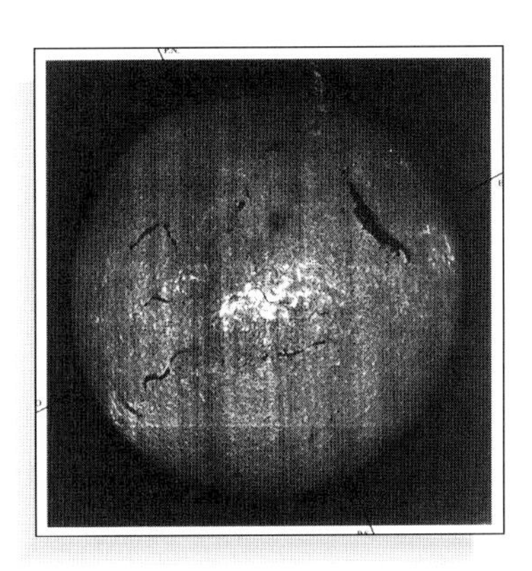

Spectroheliogram in the light of the H-α line. (Image reproduced by courtesy of the Royal Astronomical Society.)

spectrometer

A scanning *spectroscope*, little-used in astronomy.

spectrophotometry

The most detailed level of study of a *spectrum* in which the shapes of *spectrum lines* are studied and modeled to derive very detailed information on the source. It usually requires high *dispersion*, high *signal to noise* ratio spectra, and complex computer *modeling*.

spectroscope

An instrument that produces a *spectrum* of an object. Spectroscopes are used over all parts of the spectrum, but outside the *optical region* they are mostly detectors that are intrinsically sensitive to different wavelengths. In the optical region, most astronomical spectroscopes are based upon *diffraction gratings*, but *prisms* may also be used. The main components of a spectroscope are the *entrance slit*, a *collimator* to produce a parallel beam of light, the dispersing element, a lens or mirror to image the spectrum and the detector. A spectroscope that records the image by *photography* may sometimes be called a spectrograph, but most modern instruments now use *CCDs*. See also: *multi-object spectroscope*.

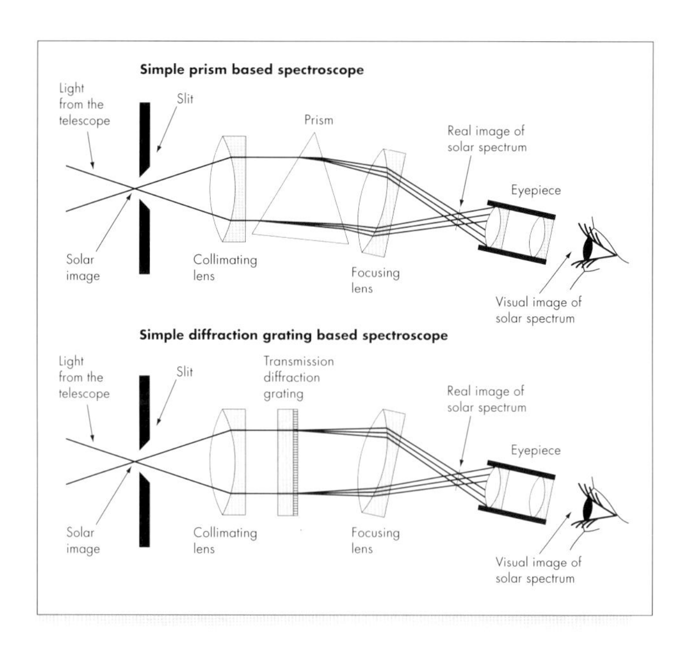

Light paths within prism and diffraction grating-based spectroscopes.

spectroscope slit

The entrance aperture to a *spectroscope*. The *spectrum* is formed from overlapping images of the slit at adjacent wavelengths, so the narrower the slit, the purer the spectrum. However, a narrow slit allows little light to enter the spectroscope, so a compromise has to be reached. See also: *stop*.

spectroscopy

The science of studying the spectra of objects. Since many processes occurring in and around stars, planets, nebulas, galaxies, etc. affect the details of a spectrum, the spectrum potentially carries vast amounts of information on those objects. For example, details of the composition, velocity, temperature, turbulence, rotation, expansion or contraction, magnetic fields and much else can be obtained from detailed studies of spectra. See also: *Kirchhoff's laws*.

spectroscopy, multi-object

See *multi-object spectroscopy*.

spectrum

A section of *electromagnetic radiation* that is arranged in order of increasing or decreasing wavelength. A hot solid, liquid or dense gas emits a *continuous spectrum*. A hot thin gas emits an *emission line* spectrum, while a hot thin gas silhouetted against a source of a continuous spectrum produces an *absorption line* spectrum. Most stars, planets and galaxies have absorption line spectra, while gaseous nebulas have emission line spectra.

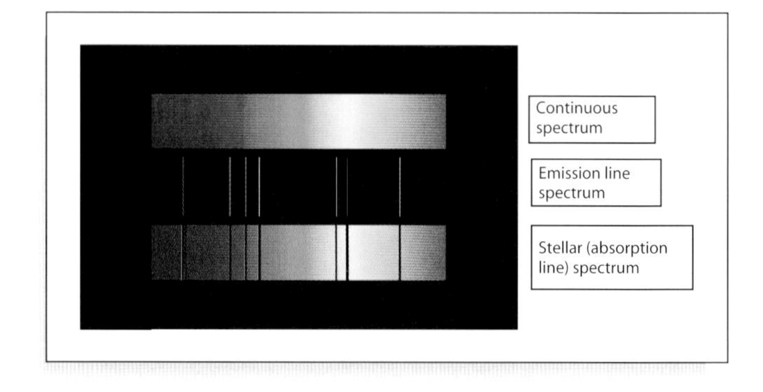

Types of spectra.

spectrum line

A bright or dark line appearing in the *spectrum* of an object. The line arises from the absorption or emission of photons at a specific *wavelength* by ions, atoms or molecules. See also: *absorption line; emission line.*

spectrum line intensity

See *equivalent width.*

speculum

An obsolete name for a *mirror*, derived from the use of *speculum metal* to produce mirrors in the past.

speculum metal

A metal alloy used in the past to make telescope mirrors, but now obsolete. Its composition was about 75% copper and 25% tin, but sometimes with small additions of arsenic and zinc. The *blank* formed from speculum metal was *ground* and *polished* by the same techniques as for *metal-on-glass mirrors*, but the reflecting surface was the actual surface of the *figured* blank. Speculum metal reflects only about 60% of the incoming light, and much less once tarnished. In order to clean a speculum metal mirror, the figured surface had thus to be re-polished and then re-figured; a lengthy process. The largest telescope to use a speculum metal mirror was Lord Rosse's 1.8m instrument at Birr castle in Ireland, constructed in 1845.

speed

See *f-ratio* and *photographic speed.*

spherical aberration

A fault in images produced by optical systems that use *lenses* or *mirrors*. It is that the optical system has a different *focal length* for *paraxial rays* compared with rays near the outer edge of the system. Spherical aberration can be eliminated for on-axis images by using a *parabolic mirror*, but is still present in off-axis images. It cannot be eliminated in a simple lens, but is minimized in a *biconvex lens*, one surface of which has three times the radius of curvature of the other. Combinations of lenses can though give high levels of correction of spherical aberration. See also: *aberration (optical).*

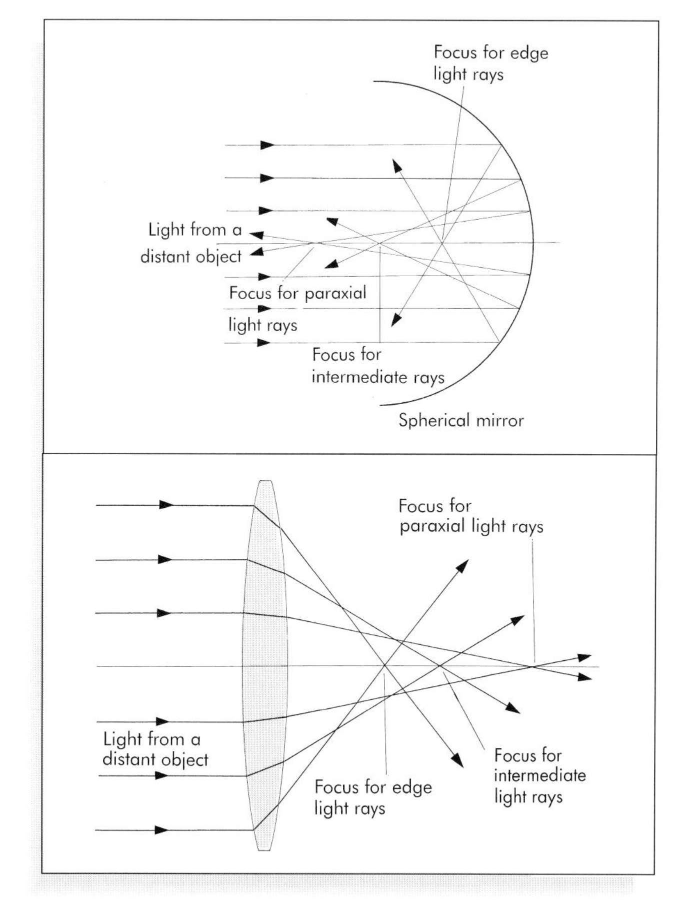

a Spherical aberration for a mirror.

b Spherical aberration for a lens.

S

spherical trigonometry

The geometry of the surface of a sphere, used to relate positions, etc. of objects on the celestial sphere. It is based upon great circles only, and for a triangle formed of three such great circles, the sides are of angular lengths, a, b, and c. The internal angle of the triangle opposite to side, a, is labelled, A, with B and C being the internal angles opposite sides b and c. There are then four formulas (only three of which are independent) that are used to interrelate these quantities:

Cosine formula: $\cos a = \cos b \cos c + \sin b \sin c \cos A$

Sine formula: $\dfrac{\sin A}{\sin a} = \dfrac{\sin B}{\sin b} = \dfrac{\sin C}{\sin C}$

Third formula: $\sin a \cos B = \cos b \sin c - \sin b \cos c \cos A$

Four parts formula: $\cos a \cos C = \sin a \cot b - \sin C \cot B$.

spider

The support arms for a *secondary mirror* in *Newtonian*, *Cassegrain* and similar telescope designs. Usually it is in the form of a cross, but sometimes intersecting circles are used to reduce *diffraction* effects.

spill over

Noise resulting from the reception by the *feed* of a *radio telescope* of direct signals (i.e. signals that have not been reflected from the main dish).

spin casting

A technique used for producing some large mirrors for telescopes. The material for the mirror is melted in a furnace that can spin on a vertical axis. The molten material therefore takes up a concave surface, and if the spin is maintained as the furnace is cooled, that shape will be retained in the final *mirror blank*. There is therefore much less material to be removed when *grinding* the mirror, especially for mirrors with small *focal ratios*. See also: *liquid mirror*.

The rotating furnace used for spin casting mirrors at the Steward observatory. (Image reproduced by courtesy of the Steward observatory mirror Laboratory, University of Arizona.)

spiral scan

See *scanning*.

spot diagram

See *ray tracing*.

spring equinox

See *First Point of Aries*.

Springfield mounting

A variant of the *Coudé mounting* for a *Newtonian telescope*. The *secondary mirror* is centered on the *declination axis* and reflects the light through the hollow axis to the tertiary mirror centered in the *polar axis*. The latter reflects the light upwards so that the observer has a convenient fixed observing position looking down the polar axis. A large counterweight is required, making the design very heavy and cumbersome.

spun-cast mirror

A mirror produce by *spin casting*. It is usually also of *honeycomb* construction to reduce its weight. See also: *liquid mirror.*

sputtering

The phenomenon upon which *aluminizing* mirrors depends. A thin aluminum wire is heated in a vacuum to below its melting point. Atoms nonetheless still evaporate from the surface, and are deposited onto the mirror.

stabilized binoculars

See *binoculars.*

standard candle method

A method for determining the distances of stars. It is based upon knowing the star's *absolute magnitude* by some method that does not involve knowing the distance (i.e. the standard candle – for example, the Cepheid period–luminosity relationship), measuring the *apparent magnitude*, and then using *Pogson's equation* to obtain the distance. See also: *distance modulus.*

standard deviation

A measure of the *uncertainty* of a set of measurements, symbol, σ. An individual measurement has a 68% chance of being closer to the mean value of the set than the standard deviation. Its value is given by:

$$\sigma = \sqrt{\frac{\sum_{i=1}^{n} (\bar{x} - x_i)^2}{n-1}},$$

where x_i is the ith of n measurements. See also: *probable error; standard error of the mean.*

standard epoch

The *epoch* used by current *star catalogs*, etc. It is usually the nearest multiple of 25 years. Thus the standard epoch now is midday 1 January 2000, the next will be midday 1 January 2025.

standard error of the mean

A measure of the *uncertainty* of the average of a set of measurements, symbol, S. Its value for n measurements is given by:

$$S = \frac{\sigma}{\sqrt{n}},$$

where s is the *standard deviation* of the set of measurements. The standard error of the mean is the uncertainty that should be quoted when the results of a set of measurements are given as the average value plus the uncertainty (i.e. $\bar{x} \pm S$).

standard star

A star whose properties represent a standard against which other stars may be compared. For example a non-variable star whose apparent magnitude is already known may act as the standard for determining the apparent magnitudes of other stars. For many purposes (UBV photometry, MKK spectral types, etc.), lists of recognized standard stars are published.

standard time

See *time zone.*

S

STAR
Simultaneous Track And Record. An *autoguiding* system for the *Starlight Xpress CCD* cameras.

star atlas
See *star chart.*

star catalog
A systematic list of the properties of a number of stars. Catalogs usually contain the star's identification and position and then list various properties such as *apparent* and *absolute magnitude*, *spectral type*, distance, type of variable, etc. They can range from specialist catalogs with just tens or hundreds of entries to massive lists like the *guide star catalog* with 19 million entries. See individual entries for further details.

star chart
A plot showing the positions, brightnesses and usually, the nature of objects in the sky. They are a convenient way of recognizing objects in the sky. They can range from commercially available products covering the whole sky with a *limiting magnitude* of the faintest stars visible to the naked eye, to the specialist *finder charts* that an observer would produce for him/herself to find a specific faint star, asteroid, comet or galaxy, etc.

Star charts and catalogs.

star chart program
See *planetarium program.* See also: *orrery software.*

star diagonal
A device incorporating a plane mirror set at 45°, or a 90° prism, that reflects light though 90°. An *eyepiece* is mounted into the star diagonal, and the diagonal in turn mounted into the eyepiece holder. Its purpose is to provide a more convenient viewing position, especially for objects around the *zenith.*

S

A star diagonal and giant eyepiece adaptor, with in the foreground an erecting prism for terrestrial observing. (Image reproduced by courtesy of Meade Instrument Corp.)

star hopping
See *finding.*

Stark effect
The splitting of spectrum lines when the atoms emitting or absorbing them are in the presence of an electric field. The effect is not of much significance in astronomy because the high conductivity of the plasmas forming stars and much of the

material in interstellar space, means that substantial voltages cannot be built up. See also: *Zeeman effect.*

Starlight Xpress
A commercial firm manufacturing *CCD* cameras for astronomical use.

Starlink
A distributed computer system used in UK universities and research institutes for the processing and analysing of data from major telescopes. Some Starlink programs are available for non-professional use.

star map
See *star chart.*

statistical parallax
The average distance to a group of stars, determined from via *proper motions.*

Stefan-Boltzmann law
See *black body temperature.*

stellar interferometer
See *Michelson stellar interferometer.*

stellar magnitude
See *absolute magnitude* and *apparent magnitude.*

stellar nomenclature
A system for identifying individual stars. For the brighter stars, the *Bayer name* or the *Flamsteed number* are often used. Their numbers in catalogs such as the *bright star catalog*, the *Henry Draper catalog* or the *guide star catalog* identify fainter stars. See also: *variable star nomenclature.*

step method
See *fractional estimate.* See also: *Pogson step method; Argelander step method.*

steradian
See *solid angle.*

stereo comparator
See *blink comparator.*

Steward observatory
An optical observatory based in Tucson, Arizona, that operates telescopes on several sites including *Kitt Peak* and *Mount Hopkins.*

STJ
See *superconducting tunnel junction detector.*

Stokes' parameters
Four parameters that describe the properties of *polarized radiation.*

Stoneyhurst solar disks
A set of transparent overlays showing the lines of solar latitude and longitude for different inclinations of the solar rotational axis to the line of sight. The position of a

sunspot, etc. can thus be measured directly by placing the appropriate overlay for the Sun's inclination on top of the solar image. See also: *Carrington rotation.*

stop

A (usually) circular aperture within an optical system. Its function is to reduce *aberrations*, or to limit background light. In camera lenses, the stop is variable in size and adjusting it allows control over the *depth of focus* and exposure length. See also: *f-ratio; spectroscope slit; iris diaphragm; baffle.*

stop down

The technique of reducing the effective size of a telescope by covering its entrance aperture with an opaque screen that has a hole smaller than the telescope's normal aperture cut into it. It may be needed for *solar observing* or to reduce the effects of *aberrations.*

Strehl ratio

A measure of the degree to which an optical system approaches (diffraction-limited) perfection. It is the ratio of the actual intensity at the center of the image to the theoretical intensity of a perfect diffraction limited image of the same object. Because of atmospheric scintillation, large Earth-based telescopes usually have Strehl ratios under 0.1, but *adaptive optics* systems may reach a value of 0.6. See also: *concentration.*

string method

A computer-based method for finding a periodic signal in noisy data. A guess is made at the period. The data is re-ordered into a single cycle on the basis of that guess and the linear distances between adjacent points added up (this would be the length of a piece of string that wound around all the points). The process is repeated with a slightly different guess at the period. When a minimum length for the 'string' is found, that guess at the period is (probably) the correct one. See also: *Fourier transforms.*

Strömgren photometry

A widely used *intermediate band photometric system.* The *filters* are centered on wavelengths, 350 nm, 410 nm, 470 nm, and 550 nm and are labelled, u, v, b, and y, respectively. Often a measurement of the hydrogen H-β line strength is added, using filters of different bandwidths and centered on 486 nm. (Standard stars for uvbyβ photometry are tabulated annually in the Astronomical *Almanac.*)

S

STScI

See *Space Telescope Science Institute.*

STScI digital sky survey

A digital version of the photographic *Palomar observatory sky survey* and the *UK Schmidt telescope* survey currently under preparation. See http://www-gsss.stsci.edu/dss/dss.html.

Student's t test

A statistical test of the probability that a measured quantity has not changed between two sets of measurements. See also: *significance level.*

style

The shadow-casting edge of the *gnomon* of a *sundial.*

Subaru telescope

An 8.2m optical telescope on Mauna Kea, Hawaii. It uses a monolithic thin mirror held in shape by *active supports*. See homepage at http://www.naoj.org/Introduction/outline.html.

The Subaru telescope. (Image reproduced by courtesy of the National Astronomical Observatory of Japan, © Subaru Telescope.)

sub-millimeter wave

A form of *electromagnetic radiation*, lying between the *radio* and *infrared* regions. The wavelength ranges from 0.3 mm to 1 mm and the frequency from 3×10^{11} to 10^{12} Hz.

summer solstice

See *solstice.*

summer time

See *civil time.*

sun dial

A device for giving *solar time* using the shadow cast from a vertical rod (the *gnomon*) onto a graduated plate. The plate is usually horizontal, but versions of sundials for use vertically on walls may be devised, and can sometimes still be seen on old churches, etc. Versions of sundials with a curved or adjustable gnomon can be constructed to read *local mean solar time*. See also: *style.*

super-apochromat

See *apochromat.*

superconducting tunnel junction detector

The superconducting tunnel junction detector (STJ) can operate out to longer wavelengths than the *CCD*, can detect individual photons, and provide intrinsic *spectral resolution* of perhaps 500 or 1000 in the visible. Instruments based upon STJs are just coming into use at the time of writing (for example the S Cam on the *William Herschel telescope*) and they seem likely to replace CCDs for many purposes in the near future.

The STJ operating principle is based upon a Josephson junction. This has two superconducting layers separated by a very thin insulating layer. Quantum mechanical effects allow the electrons to tunnel across the junction, and so a current may flow across the junction despite the presence of the insulating layer. That current may be suppressed by an externally applied magnetic field. The STJ detector therefore comprises a Josephson junction placed within a magnetic field to suppress the current, and with an electric field across it. An incident photon may split one of the pairs of electrons within the superconductor, and these free electrons are able to tunnel across the junction under the influence of the electric field,

S

producing a detectable burst of current. The energy required to split the electron pairs is very small compared with that needed to produce an *electron-hole pair* within a CCD. STJs therefore can potentially detect millimetre-wavelength photons, and at shorter wavelengths the current induced by a photon will be proportional to that photon's energy, so providing the detector's intrinsic spectral resolution.

super heterodyne receiver

See *radio receiver.*

superior conjunction

A *conjunction* between the Sun and Mercury or Venus, when the planet is farthest from the Earth. Although the whole of the illuminated surface of the planet is then visible, it is not the best occasion for observation (see *greatest elongation*) since the angular size of the planet is at its smallest. See also: *aspect.*

super Kamiokande

See *Kamiokande neutrino detector.*

super Plössl eyepiece

See *Plössl eyepiece.*

super Schmidt camera

A variant of the *Schmidt camera* that has additional *correcting lenses* giving it *focal ratios* as short as f0.5. Used especially for photographing meteors.

surface brightness

The brightness per unit area of an *extended object* such as a galaxy. It is usually expressed as magnitudes per square second of arc (i.e. the *apparent magnitude* that a 1 second of arc square of the object would have in isolation). If an extended object is of uniform brightness, and of area, n square seconds of arc, then its surface brightness, S, is given by $S = m_{int} + 2.5 \log_{10} n$, where m_{int} is the *integrated magnitude* of the object. Thus the full moon (integrated magnitude, -12.7^m) has a surface brightness of 3.3 magnitudes per square second of arc, while the Ring nebula (M57, integrated magnitude, 9.4^m) has a surface brightness of about 19 magnitudes per square second of arc.

Surface brightness may also be expressed in watts per square meter per steradian, when it is likely to be called the radiance of the object. See also: *sky background.*

S

sweet spot

A combination of *telescope* and *eyepiece* that gives images of maximum acuity. It varies somewhat from one observer to another, but typically occurs with an eyepiece whose *focal length* in millimeters is twice the *effective focal ratio* of the telescope (e.g. a 20mm eyepiece on an f10 telescope).

synchrotron radiation

Electromagnetic radiation emitted by high velocity charged particles, usually electrons, that are moving along a curved track due to a magnetic field. It can be emitted at any wavelength, but for most astronomical sources it is to be found in the *radio region.* It is strongly *polarized* and has a *spectral index* usually between –0.5 and –1.5. See also: *free–free radiation.*

synodic month

The average interval between recurrences of the same lunar phase, also known as the lunation. Its duration is 29.531 days.

synodic period

The time interval required for a planet, etc. to return to the same relative position with respect to the Earth. For example the length of time between successive *oppositions* of an outer planet. It is related to the *sidereal period*, P, by:

$$S = \frac{P}{1 - P}$$

for an inner planet and by:

$$S = \frac{P}{P - 1}$$

for an outer planet, where S is the synodic period and S and P are in units of years. Thus the sidereal period of Venus is 224.7 days (0.615 years) so its synodic period is 583.9 days (1.60 years).

synodic rotation period

The rotation period of a body as seen from the moving Earth. For example the sidereal rotation period of the Sun at 20° solar latitude is 25.38 days, whereas from the Earth it appears to be 27.28 days.

synthetic aperture radar

A *radar* system on board a spacecraft that operates in a manner similar to an *aperture synthesis* interferometer in order to obtain higher resolution. Synthetic aperture radar (SAR) uses a single antenna that is moved by the spacecraft's motion to occupy the successive positions that the multiple antennas of an aperture synthesis system would have. Recording and applying appropriate time delays allows the high-resolution image to be synthesized. The linear resolution of a SAR is half the diameter of the radar dish. Uniquely therefore, the resolution of a SAR is improved by decreasing the size of its dish. However, since this will reduce the system's sensitivity, a compromise between the two requirements always has to be reached.

systematic noise

Noise that has a trend or bias to its effects, i.e. it is not random. For example, *atmospheric extinction* or changes to the response of electronic equipment with temperature. Often the effects of systematic noise can be predicted and thus eliminated during *data reduction*. See also: *random noise; white noise.*

T

TAA
Track And Accumulate. An *autoguiding* system for the *SBIG CCD* cameras.

T-adaptor
An adaptor to fit most *digital* and 35mm *SLR cameras* to enable different lenses to be used or the camera to be attached to a telescope.

TAI
See *international atomic time.*

Takahashi
A commercial firm manufacturing telescopes and accessories.

Tal
A commercial firm manufacturing telescopes and accessories.

TAMA 300
See *gravity wave detector.*

tapering
See *antenna tapering.*

Tartu observatory
An optical observatory near Tartu in Estonia.

TDI
See *tracking.*

tear-drop effect
An effect that is noticeable during *transits* of Mercury and Venus, or with other very high contrast images. At the start (or end) of the transit, the black disk of the planet silhouetted against the solar photosphere seems to remain connected to the edge of the Sun by a dark lane, making it appear like a tear-drop, for a short while after it has actually crossed completely over the edge. The effect arises from cross-linkages between the rods and cones in the eye's *retina*. See also: *irradiation.*

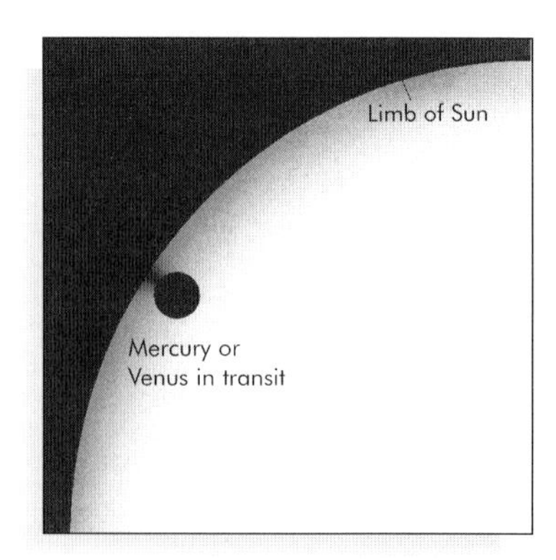

The tear drop effect.

Technical Innovations
A commercial firm manufacturing telescope domes.

Teflon
A plastic (PTFE) material with a very low coefficient of friction. It is often used to form the bearing surfaces for *Dobsonian telescope mountings*.

Teide observatory
An optical observatory on Mount Teide, Tenerife, altitude 2400 m.

tele-compressor
A *positive lens* placed before the *eyepiece* or *detector* to reduce the *image scale* and so to provide a wider *field of view*.

tele-extender
See *Barlow lens*.

telephoto advantage
The advantage that some telescope designs, such as the *Schmidt–Cassegrain* and *Maksutov*, possess of having a longer *effective focal length* than the physical size of the instrument, through the use of a beam-expanding *secondary mirror*.

telescope
An instrument that gathers *electromagnetic radiation* from a small area of the sky and concentrates it into an image of the objects within that area. The term is most widely used for instruments operating in the *optical* and *radio regions*. In the optical region, telescopes are based upon lenses (*refractors*) or mirrors (*reflectors*) or a combination of both (*catadioptric telescopes*). See individual entries for the designs of optical telescopes. *Radio telescopes* are all based upon reflection. In the X-ray region, telescopes are based upon *grazing incidence* reflection.

The term is also often used for instruments which are not strictly telescopes and that detect *gamma rays*, *cosmic rays*, *neutrinos* or *gravity waves*. These instruments are more correctly called detectors since they do not concentrate the particles or radiation, and in some cases are not direction-sensitive.

telescope, invention of
The invention of the optical telescope is generally attributed to a Dutch spectacle maker named Hans Lippershey in 1608. However, it is not known what the optical design of his instrument was. Galileo, after hearing of Lippershey's instrument, invented the *Galilean refractor* in 1609. The reflecting telescope was invented in 1663 by James Gregory (see *Gregorian telescope*), but the first reflector to be built was produced by Newton in 1668. See also: *Digges' telescope; Newtonian telescope*.

telescope mounting
See *alt-azimuth telescope mounting* and *equatorial telescope mounting*. See also: *wedge*.

telescope tube
That part of a telescope, which holds the optical components and keeps them in their correct mutual alignment whilst they are pointed to different parts of the sky. In refractors and small reflectors it may literally be a tube, but in larger instruments it is usually a skeleton structure formed of cross-braced struts. If needed to reduce background light, the spaces between the struts may be filled-in with non-structural sheeting. See also: *Serrurier truss*.

television camera

See *video camera.*

Televue

A commercial firm manufacturing telescopes and accessories.

telluric

Appertaining to the Earth.

Telrad finder

A commercially available *red dot finder.*

terminator

The dividing line between the illuminated and shadowed sections of the Moon, a planet or a planetary satellite.

terrestrial dynamical time

An old name for terrestrial time – see *ephemeris time.*

terrestrial telescope

A telescope producing an *erect image* and therefore suited to observing objects on the Earth. *Galilean* and *Gregorian telescopes* provide erect images directly; other designs need a *relay lens* or other additional optical components to reverse their normally *inverted images*. See also: *astronomical telescope.*

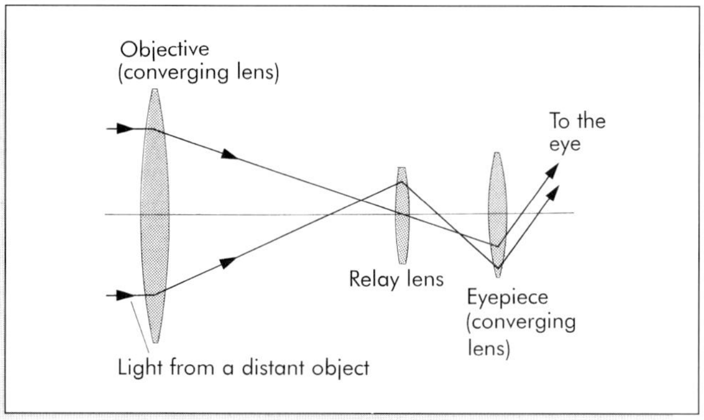

Light paths within a terrestrial telescope.

terrestrial time

See *ephemeris time.*

thermal broadening

See *Doppler broadening.* See also: *line broadening.*

thermal noise

See *noise.*

thermal radiation

Electromagnetic radiation emitted by an object at a temperature above absolute zero. For many objects, including stars, the emission is similar to that of a *black body* and follows the Planck distribution, given by:

$$I_\lambda = \frac{2hc^2\mu^2}{\lambda^5(e^{hc/\lambda kT}-1)},$$

where I_λ is the intensity at wavelength, λ, per unit wavelength interval, h is Planck's constant ($h = 6.626 \times 10^{-34}$ J s), c is the velocity of light in a vacuum ($c = 2.998 \times 10^8$ m s^{-1}), μ is the refractive index of the medium (normally close to 1), k is Boltzmann's constant ($k = 1.381 \times 10^{-23}$ J K^{-1}) and T is the temperature. The total flux, F, increases with temperature and is given by $F = \sigma\mu^2T^4$. This is known as Stefan's law and σ is Stefan's constant ($\sigma = 5.670 \times 10^{-8}$ W m^{-2} K^{-4}). The wavelengths over which the main emission occurs decrease as the temperature increases. Thus we see cool stars as reddish, hotter ones as yellow, and very hot stars as blue-white (see Wien's law in the *black body temperature* entry).

Free-free radiation arising from the thermal motions of electrons in hot plasmas is also usually included under the heading of thermal radiation. See also: *black body temperature; Rayleigh-Jeans approximation.*

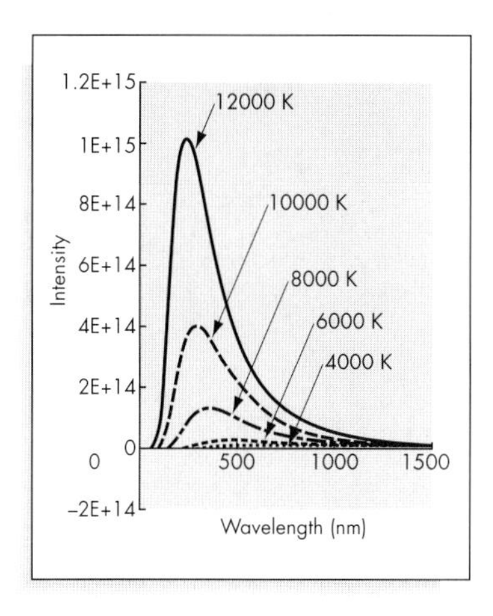

Thermal radiation and its variation with wavelength and temperature.

thermal stress

Stress induced into an object because different parts of the object are at different temperatures and are therefore expanding or contracting by different amounts. It is most important in astronomy for large telescope *mirrors*, where the stresses may distort the optical surface. Such mirrors are now therefore made from low thermal expansion materials such as *Cer-Vit* and *ULE*. The mirror may also be cooled or heated to the predicted nighttime temperature in advance of an observing session to reduce the effect further.

thermocouple

A detector of *electromagnetic radiation* whose operating principle is based upon the Seebeck effect. That effect is that two dissimilar metals in contact with each other develop a voltage across their junction whose magnitude varies with the temperature. A simple thermocouple is thus made from two thin wires, usually of antimony and bismuth, that are joined together in a circuit together with a sensitive ammeter. If one junction is kept at a constant temperature, while the other is heated by the radiation, then the voltages at the two junctions will differ and a current will flow whose magnitude is proportional to the intensity of the radiation. Thermocouples are not very sensitive detectors, but since they rely only on the heating effect of the radiation, they can be used from the *microwave region* to the *ultraviolet* by coating then with an appropriate absorber. They can therefore be used to calibrate and compare detectors of greater sensitivity but whose wavelength range is more restricted.

thermopile

A detector of *electromagnetic radiation* that consists of several *thermocouples* joined together.

thin mirror

A monolithic mirror, especially the *primary mirror* of a large telescope that is too thin to retain its correct shape under its own weight. The correct shape has to be maintained by using *active supports*. See also: *honeycomb mirror; segmented mirror; spun-cast mirror.*

thinned CCD

See *CCD detector.*

threshold adjustment

See *grey scaling.*

throughput

A measure of the efficiency of an optical system. It is the amount of energy passed by the system when it is illuminated by unit intensity per unit area and per unit solid angle.

TIFF

See *electronic image format.*

tilt correction

See *tip-tilt mirror.*

time

See *civil time*, *coordinated universal time*, and *international atomic time.*

time constant

The speed of response of a detector or other device to a change in its input.

time delay integration

See *tracking.*

time zone

One of 24 regions of the Earth, across each one of which the use of a standard time has been agreed. The zones are mostly 15° of longitude wide, starting with the zone centered on the *Greenwich meridian*, but some variations occur so that the time used across individual nations is uniform.

tip-tilt mirror

A device used to give partial correction of atmospheric *scintillation* (see *adaptive optics*). It is a plane mirror that can be tilted rapidly in any direction to compensate for the overall inclination of the wave front caused by the atmospheric delays. It can produce a significant improvement in image sharpness at a much lower cost that a full adaptive optics system. Commercially produced devices are now available for use on small telescopes.

T-mount

See *T-adaptor.*

tool

A disk of glass used in the production of a *mirror* – see *grinding.*

top hat exposure

A method of obtaining images on small telescopes with a minimum of shake. It is used for *exposure* lengths of a second to a minute or so. The method is to set up the telescope and camera, and then to hand hold a cover over the telescope objective. The cover is sufficiently close to the telescope to prevent light entering, but not actually touching it (the name of the method derives from top hats being used for the purpose in the past). The camera shutter is then opened and held open. A few seconds is allowed to elapse so that vibrations die away, and then the exposure made by moving the cover off and then back over the telescope. The camera shutter is then closed.

topocentric coordinates
Coordinates (e.g. *right ascension* and *declination*) as they would be seen from a specific point on the surface of the Earth. See also: *geocentric coordinates; heliocentric coordinates.*

Torun observatory
A radio observatory with a 32m dish at near Torun in Poland.

total magnitude
A synonym for *integrated magnitude.*

tower solar telescope
A *solar telescope* that has its objective mounted at the top of a high tower in order to be above the ground turbulence. (CAUTION – always take appropriate precautions when observing the Sun or damage to the eye and/or telescope can result.)

The solar tower telescope at Mount Wilson.

tracking
The movement of a telescope in order to keep it accurately aligned on an object in the sky. For a telescope on an *equatorial mounting* this just requires a rotation around the *polar axis* at a rate of 360° per *sidereal day*. For telescopes on *alt-azimuth mountings*, both axes must be driven at variable rates, and this is now accomplished using computers to control the motors.

If only small changes are needed, then tracking may be undertaken by moving the charges from *pixel* to pixel within a *CCD* to follow the movement of the image over its surface. This technique is sometimes known as time delay integration (TDI). See also: *guiding.*

trailed image
The image of an object when either the telescope has not *tracked* the object accurately, or the object has moved during the exposure. Star images are thus short lines instead of circular disks. A trailed image is usually a fault, but may be deliberate on some occasions. Thus for example, *guiding* on a comet, will result in all the background stars' images being trailed because of the comet's motion, but the image of the comet will be sharp.

Trailed images of stars resulting from guiding on the comet (comet Mueller 1993a). (Image reproduced by courtesy of Bob Forrest, University of Hertfordshire Observatory.)

trailing edge
The rear edge of a moving extended object like the Sun or Moon.

trails
A line across an image arising from the lights of an aircraft flying through the telescope beam during the exposure, or similarly from the reflected light from a spacecraft.

transfer function
See *modulation transfer function.*

transfer lens
A lens within an optical system whose sole purpose is to move the focus to a convenient position. See also: *relay lens.*

transit
1 The passage of one object in the sky in front of another, when the nearer object has a much smaller angular size than the more distant one. For example, the transit of Venus across the Sun. (CAUTION – always take appropriate precautions when observing the Sun or damage to the eye and/or telescope can result.) See also: *eclipse; occultation.*
2 The passage of an object across the *prime meridian.* See also: *culmination.*
3 The passage of a feature on the surface of a planet, the Sun, the Moon, etc. across the center line of that body's disk as seen from the Earth.

transit circle
A telescope that is mounted on a fixed east–west axis, and so can only be used to observe objects crossing the *prime meridian.* It is used to measure the altitude of objects on the prime meridian (and so determine their declination), when it is called a transit telescope, and for timing the meridian passage of objects in order to calibrate clocks, when it is called a meridian circle.

transition
An energy change occurring within an atom, ion or molecule that leads to the emission or absorption of a photon. Most often it is visualized as an electron moving from one 'orbit' to another, but it may also arise through a molecule changing its rate of vibration or rotation.

transition probability
The probability of the occurrence of a transition. There are three transition probabilities for any transition; those for absorption, spontaneous emission and stimulated emission (where a second photon of the same wavelength prompts the downward transition). The values of transition probabilities range from 10^8 s^{-1} for normal transitions to 10^3 s^{-1} to 10^{-15} s^{-1} or less for *forbidden lines.* See also: *lifetime.*

transit telescope
See *transit circle.*

transmission grating
A *diffraction grating* where the individual apertures are long thin slits. See also: *blazed grating; replica grating; reflection grating.*

transmittance

A measure of the ability of a semi-transparent medium to transmit radiation. It is given by the ratio of the intensity of the incident radiation to that of the emergent radiation. The reciprocal of transmittance is *opacity*. See also: *optical depth; optical density.*

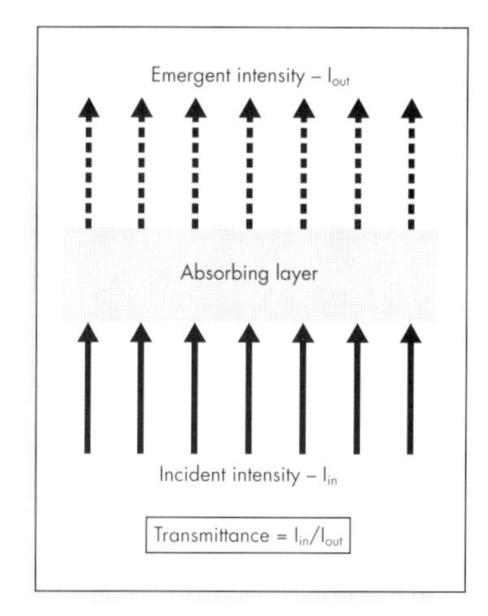

Transmittance.

trigonometrical parallax

See *parallax.*

triple loupe eyepiece

See *monocentric eyepiece.*

triplet

1 A lens formed from three components, such as the *apochromat.*
2 Three close *spectrum lines* arising from the same atom, ion or molecule and with a common *energy level.*

tripod

A portable support for a small telescope, camera, binoculars, etc. that has three legs attached to a central platform, or it may sometimes take the form of a central pillar with three horizontal buttresses. Tripods are usually made to be folded or disassembled for transport. See also: *alt-azimuth telescope mounting; equatorial telescope mounting; pier.*

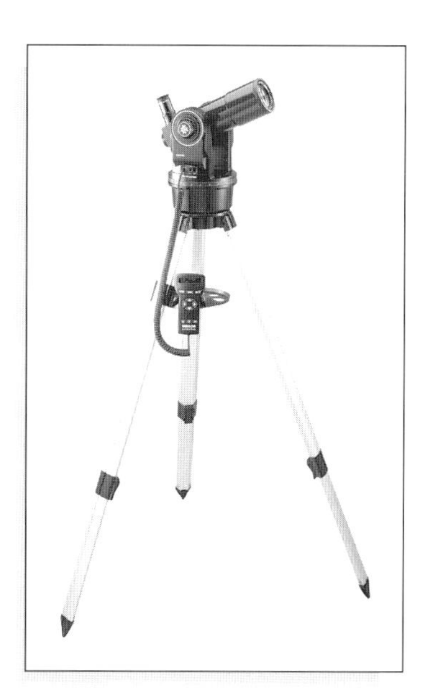

A tripod for a small telescope. (Image reproduced by courtesy of Meade Instrument Corp.)

tri-Schiefspiegler telescope

A *Schiefspiegler* design with three mutually tilted and curved mirrors arranged in such a way that the optical path is unobstructed and the aberrations introduced by the tilting mirrors are cancelled out.

tropical year

The year upon which calendars are based. It is the time for the Sun to return to the *First Point of Aries*, and is equal to 365.2422 days. It differs from the *sidereal orbital period* of the Earth (the *sidereal year*), because of the movement of the First Point of Aries due to *precession*.

true field of view

The angular diameter of the area of sky seen through a telescope. It is given by the *field of view* of the *eyepiece* (or *apparent field of view*) divided by the *magnification*.

true horizon

See *horizon*.

turbulence broadening

See *Doppler broadening*. See also: *line broadening*.

TV camera

See *video camera*.

twenty-one centimeter line

A spectrum line from the hydrogen atom at a wavelength of 21 cm (1.42 GHz). It arises as the electron in the ground state of the atom undergoes a *transition* from having its spin parallel to that of the nucleus to anti-parallel. It has a very low *transition probability* (3×10^{-15} s^{-1}), but the huge number of hydrogen atoms means that it is easily detectable, and so it is used to map out the Milky Way galaxy.

twilight

The interval after sunset, or before dawn, when light from the Sun is scattered towards the observer giving some local illumination, and some background light in the sky. Civil twilight begins and ends when the Sun is 6° below the horizon, nautical twilight when the Sun is 12° below the horizon, and astronomical twilight when the Sun is 18° below the horizon. (Tabulated annually in the Astronomical *Almanac*.)

twinkling

See *scintillation* and *seeing*.

two color diagram

A plot of the (U–B) *color index* versus the (B–V) color index for a group of stars. See also: *Hertzsprung–Russell diagram*.

two star photometer

T

A *photometer* that makes simultaneous measurements of two stars. These will usually be the star of interest and the *comparison star*, or the star of interest and the *sky background*. See also: *multi-channel photometer*.

Tycho catalog

A catalog of the positions and magnitudes of some 1,058,000 stars produced by the *Hipparcos spacecraft*. The uncertainties in its data are range from about 0.007″ to 0.025″. It may be accessed at http://archive.ast.cam.ac.uk/hipp/. See also: *Hipparcos catalog*.

U

U–B color index
See *color index.*

UBV photometric system
An extensively used *wide band photometric system*, defined by Harold Johnson and William Morgan in 1953. The *filters* are centered on wavelengths, 365 nm, 440 nm, and 550 nm and are labelled, U (ultraviolet), B (blue) and V (visual), respectively. The filters must be used in conjunction with a *photomultiplier* to give the correct response curves. If other detectors are used, then correction factors may be needed. The (U–B) and (B–V) *color indices* are frequently used to determine the temperatures of stars. The U band is badly affected by atmospheric absorption at low altitudes, and so may give variable results.
The system has been extended to longer wavelengths. The most widespread such extension has additional filters (or uses the atmospheric windows at long wavelengths) at:

R	700 nm
I	900 nm
J	1.25 μm
K	2.2 μm
L	3.4 μm
M	4.9 μm
N	10.2 μm
Q	20 μm

(Standard stars for UBVRI photometry are tabulated annually in the Astronomical *Almanac.*)

UK infrared telescope
A 3.8m telescope observing from 1 μm to 30 μm and sited at *Mauna Kea observatory.* It was the first major telescope to use a *thin mirror.*

UKIRT
See *UK infrared telescope.*

UK Schmidt telescope
A 1.2m *Schmidt camera* sited at the *Anglo Australian observatory.*

UKST
See *UK Schmidt telescope.*

ULE
A low thermal expansion glass-ceramic material, widely used to form mirrors for large telescopes (the acronym is from Ultra-Low Expansion).

ultraviolet
A short *wavelength*, high *frequency*, form of *electromagnetic radiation*, but less extreme than *X-rays* and *gamma rays.* The wavelength ranges from 350 nm to 1 nm and the frequency from 8.6×10^{14} Hz to 3×10^{17} Hz. In energy terms, ultraviolet photons have energies between 3.5 eV and 10^3 eV.

ultra-wide angle eyepiece

A modern class A (see *eyepieces*) eyepiece with up to eight elements and a *field of view* of up to 84°. The *eye relief* is about two thirds of the focal length. The design is well corrected for aberrations and is similar to the *Nagler*.

U magnitude

See *UBV photometric system*.

uncertainty

A measure of the precision of a measurement or set of measurements. No measurement can ever be made exactly, and several measurements of the same quantity will be affected by random fluctuations. In most cases the distribution of several measurements of the same quantity will show a bell-shaped variation known as the *Gaussian* or normal distribution. The narrower the bell shape, the more precise are the measurements. The width of the Gaussian curve is given by the *standard deviation*, and the uncertainty in the average of the set of measurements is given by the related *standard error of the mean*. See also: *probable error*.

under-sampling

The use of a detector with a lower resolution than that required to obtain all the information that is present in an image.

un-filled aperture

See *aperture synthesis*.

United States Naval observatory

An optical observatory founded in 1844 and based in Washington, DC. It has an observing station at Flagstaff, Arizona.

unit power finder

A synonym for *zero power finder*.

universal time

A measurement of time based upon the Earth's rotation taking exactly 24 hours. It is the *mean solar time* at the equator on the Greenwich meridian. It replaced Greenwich mean time in 1928. There are three levels of universal time, UT0 that is based directly on observations, UT1 that has corrections for the motion of the Earth's pole included and UT2 that also incorporates changes in the speed of the Earth's rotation. See also: *coordinated universal time; international atomic time.*

unsharp masking

An *image processing* technique that permits the display of all the details of an image with very wide *dynamic range*. The technique originated with *photography*, but computer-based image processing now has equivalent packages. In photographic unsharp masking, a positive contact print onto film (not printing paper) is made from the original negative. The original and the copy are then aligned with each other and separated by a short distance. Another print is made of the negative through the combination. The positive is slightly out of focus but is densest where the negative is lightest and vice versa and so acts as a filter to reduce the most intense parts of the image in comparison with the fainter sections.

upper culmination

See *culmination*.

upper transit
A synonym for upper *culmination.*

upright image
See *erect image.*

USNO
See *United States Naval observatory.*

UT
See *coordinated universal time.*

UTC
See *coordinated universal time.*

uvby photometric system
See *Strömgren photometry.*

uv plane
An imaginary plane within which one element of an *aperture synthesis* system moves around the other as the Earth rotates.

UWA eyepiece
See *ultra-wide angle eyepiece.*

U

V

variable density filter

A neutral *filter* used to reduce the brightness of the Moon (it must not be used for solar observing). It is formed from two *Polaroid* sheets whose angle may be varied so that the transmission can vary between 0 and 50%.

variable star nomenclature

There are several systems for denoting variables, and many individual stars have several designations. The extension to the *Bayer system* is the most widespread approach. This labels the variables in each constellation with capital letters, starting at R for the brightest. Letters from R to Z suffice for only nine variables in each constellation, and so thereafter double capitals are used. RR to RZ is used for the next nine variables discovered, then SS to SZ for the next eight, TT to TZ for the next seven, and so on to ZZ, giving another 54 labels in all. After that the sequences AA to AZ, BB to BZ, CC to CZ ...QQ to QZ are used. A total of 334 variables in each constellation may be thus identified, with the letters followed by the constellation abbreviation; S Cyg, RR Lyr, FG Sag, etc. When more than 334 variables are found the simpler practice of using the letter V followed by a number starting from 335 is used. Thus we have V861 Sco, V444 Cyg, and V432 Her, etc.

Pulsars and other special groups of variables may use other systems of labeling; a common practice is to use a label to indicate the variable type and combine it with an approximate right ascension and declination. Thus we have the pulsar, PSR 1257+12, which is to be found near RA 12h 57m and Dec +12°.

Vatican observatory

An optical observatory based at Castel Gandolfo near Rome, and with an out station at the *Steward observatory*, Arizona.

vernal equinox

See *First Point of Aries*.

vernier scale

A device to allow a graduated scale to be read accurately to a fraction of its smallest division. The fiducial mark for the scale is itself a small scale. If the main scale is to be subdivided into ten, then ten divisions on vernier scale have the same length as nine divisions on the main scale. The mark on the vernier scale that aligns most closely with a mark on the main scale then gives the sub-division of the reading from the main scale.

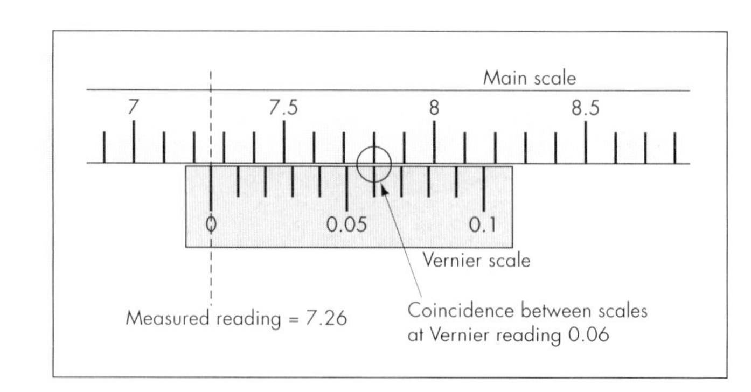

A Vernier scale.

vertex sphere

A sphere whose curvature matches that of a *paraboloid mirror* on its *optical axis*. It has a radius that is twice the *focal length* of the mirror.

vertical circle

A *great circle* on the *celestial sphere* that passes through the *zenith*.

very large array

A radio *aperture synthesis* system with 27 dishes each 25m across, sited in New Mexico. The dishes may be moved along three tracks up to 21km long arranged in a Y shape, giving a maximum base line of 36 km.

very large telescope

See panel (page 248).

very long base-line interferometry

A radio *interferometer* with base lines up to several thousand kilometers in length. Although the theoretical basis of the operation is the same as that of a smaller system (like *MERLIN*), the practical operation is slightly different. Each fixed antenna records the signal simultaneously with the others in the system and also records the pulses from an atomic clock. The recordings are then brought together and the clock pulses used to synchronize them when they are combined. Resolutions of less than a milli-arc second can be reached.

vibrational transition

See *transition*.

Victor Blanco telescope

A 4.0m optical telescope on Cerro Tololo, Chile.

video camera

A *camera* that uses *a CCD detector* to obtain moving images. The cameras used by professional TV companies, especially for studio work, may still sometimes be based on vidicon tubes.

vignetting

An uneven illumination of the *image plane* arising from shadows cast by structural elements of the instrument that intrude into the light path, and from incomplete illumination of optical components.

violet cut-off filter

A filter that eliminates the violet part of the spectrum. Since this light is often out of focus for many camera lenses, when the image is nominally in focus, the use of such a filter will provide sharper images of the night sky.

VIRGO

See *gravity wave detector*.

virtual image

An image that behaves "as if it were there", but cannot be projected onto a screen. Produced, for example, when a parallel beam of light falls onto a *diverging lens*. See also: *real image*.

visibility function

A measure of the intensity of the fringe pattern in an *interferometer*. It is a plot of the way that

$$\frac{I_{max} - I_{min}}{I_{max} + I_{min}}$$

varies with antenna spacing or source separation (where I_{max} is the intensity of a maximum in the fringe pattern, and I_{min}, the intensity of a minimum).

very large telescope

A complex of four 8.2 m *Ritchey–Chrétien* optical telescopes plus three 1 m telescopes at the *European Southern Observatory*, Cerro Paranal, Chile. The large telescopes use monolithic thin mirrors held in shape by *active supports*. The telescopes are individually named in the local (Mapuche) language after astronomical objects: Antu (Sun), Kueyen (Moon), Melipal (Southern Cross) and Yepun (Evening star – Venus). The telescopes can operate independently or together to form an *interferometer*. See homepage at http://www.eso.org/projects/vlt/.

The enclosures ('domes') for the European Southern Observatory's Very Large Telescope. (Image reproduced by courtesy of the European Southern Observatory.)

The Very Large Telescope – ANTU. (Image reproduced by courtesy of the European Southern Observatory.)

The Very Large Telescope – KUEYEN. (Image reproduced by courtesy of the European Southern Observatory.)

The Very Large Telescope – the FOcal Reducer and Spectrograph (FORS) on KUEYEN. (Image reproduced by courtesy of the European Southern Observatory.)

visible light

A form of *electromagnetic radiation*, lying between the *infrared* and *ultraviolet* regions, and to which the eye is sensitive. The wavelength ranges from 700 nm to 350 nm and the frequency from 4.3×10^{14} Hz to 8.6×10^{14} Hz. In energy terms, visible photons have energies between 1.8 eV and 3.5 eV. See also: *optical region.*

VISTA telescope

A 4m optical wide field survey telescope for the UK, to be constructed on Cerro Paranal in Chile, altitude 2600 m.

visual estimate

See *fractional estimate* and *angular estimate.*

visual magnitude

The apparent magnitude of an object estimated by eye. Its value is similar to that of the V magnitude (see *UBV photometric system*).

visual purple

See *retina.*

Vixen

A commercial firm manufacturing telescopes and accessories.

VLA

See *very large array.*

VLBI

See *very long base-line interferometry.*

VLT

See *very large telescope.*

V magnitude

See *UBV photometric system.*

V

waning
The sequence of phases of the Moon, Mercury or Venus going from *full* to *new*. See also: *waxing*.

wave band
A region of the electromagnetic spectrum between two defined wavelengths or frequencies. See also: *bandwidth*.

wave front
A surface, perpendicular to the direction of travel of a beam of *electromagnetic radiation*, that connects points within the wave that have corresponding *phases*. It is flat for a parallel beam of radiation, and concave towards the focus in a focussed beam. The quality of telescope optics is sometimes measured by the deviations from perfection that they produce in the wave front. For a reflector this criterion has twice the value of the deviations of the surface of the mirror. Thus to achieve *Rayleigh resolution*, the wave front deviations from perfection must be less than $\lambda/4$. See also: *diffraction limited optics; $\lambda/4$ criterion. $\lambda/8$ criterion.*

waveguide
A transmission line for high frequency *radio waves* consisting of a rectangular metal pipe.

wavelength
The linear distance between successive maxima (or minima) of a wave motion, usually symbolized by λ. See also: *frequency; period.*

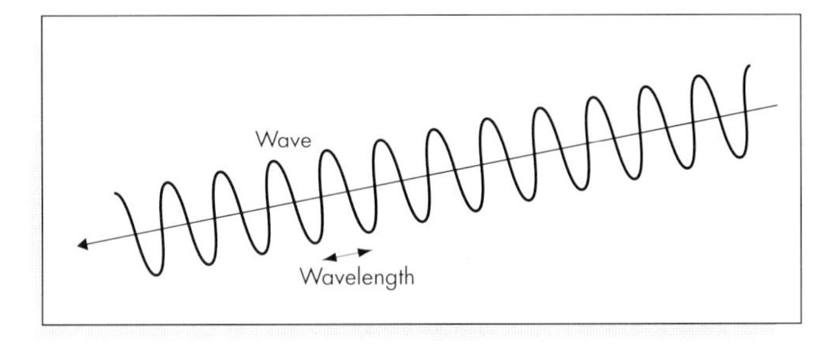

Wavelength.

wave noise
Signal *noise* arising from the quantum nature of *electromagnetic radiation.*

wave number
The number of *wavelengths* in a meter.

wave plate
A device used in *polarizers* to convert *circularly* or *elliptically polarized* light into *linearly polarized* light, or vice versa (a *quarter wave plate*), or to rotate the direction of linearly polarized light (a *half wave plate*). See also: *Pockel's cell.*

wave zone
A synonym for *far field.*

waxing
The sequence of phases of the Moon, Mercury or Venus going from *new* to *full.* See also: *waning.*

Weber–Fechner law
The rule that the response of the *eye* to variations in intensity is geometric rather than arithmetic. That is, if the eye perceives stars A and B to differ by an amount similar to the difference between stars C and D, then it will be the RATIO of the energy from A to that from B that will be the similar to the RATIO of the energy from C to that from D, and not their energy DIFFERENCES. This effect has given rise to the form of the *Pogson equation* for stellar *apparent magnitudes,* since that is based upon eye estimates of stars' brightnesses.

wedge
A device that may be fitted to a small *alt-azimuth* telescope to convert it to an *equatorially* mounted telescope. The wedge has a mounting plate that fits onto the base of the telescope and whose angle to the horizontal may be adjusted. The mounting plate is adjusted to an angle equal to 90° minus the observer's latitude, and then swiveled until the telescope's azimuth axis is aligned on the *celestial pole.* The azimuth axis then becomes the *polar axis* and the altitude axis becomes the *declination axis.*

well capacity
The number of electrons that a single *CCD* element can hold before *saturation.*

Westerbork observatory
A radio observatory at Westerbork in the Netherlands, with an aperture synthesis system comprising fourteen 25m dishes and a maximum baseline of 3 km.

western elongation
The *elongation* when an object is to the west of the Sun (i.e. it sets before the Sun). See also: *eastern elongation; greatest elongation; aspect.*

white noise
Noise that has the same intensity at all frequencies (i.e. like white light, it covers all frequencies). See also: *random noise.*

WHT
See *William Herschel telescope.*

wide band filter
A *band-pass filter* with a broad *bandwidth.* Typically the bandwidth will be 20 to 100 nanometers. See also: *filter; UBV photometry.*

W

wide band photometry
Photometry in the *optical region* that uses filters with bandwidths of 50 to 100 nm or more. See for example the *UBV photometric system.*

wide field eyepiece
See *eyepiece.*

wide field telescope

A telescope optimized to give a wide *field of view*. It requires the use of a wide-angle *eyepiece*, a smallish *objective* and the *minimum magnification*. Fields of view of 2° to 6° may then be obtained. See also: *binoculars*.

widener

A device used in a *spectroscope* to widen the *spectrum* of a star, so that it is easier to measure. It often consists of a thick block of glass with parallel sides that is placed in the optical path before the detector and rocked from side to side. The spectrum is thus displaced up and down the detector, and a wider image produced.

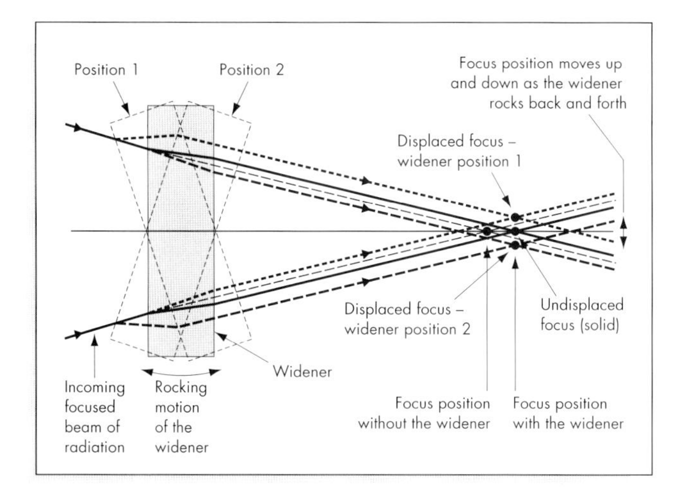

A rocking widener for a spectroscope.

Wiener filter

A mathematical operation applied to the *Fourier transform* of a set of data that reduces the effects of *random noise*.

Wien's law

See *black body temperature*.

William Herschel telescope

A 4.2m *Ritchey–Chrétien* optical telescope with a monolithic mirror on La Palma, Canary Islands.

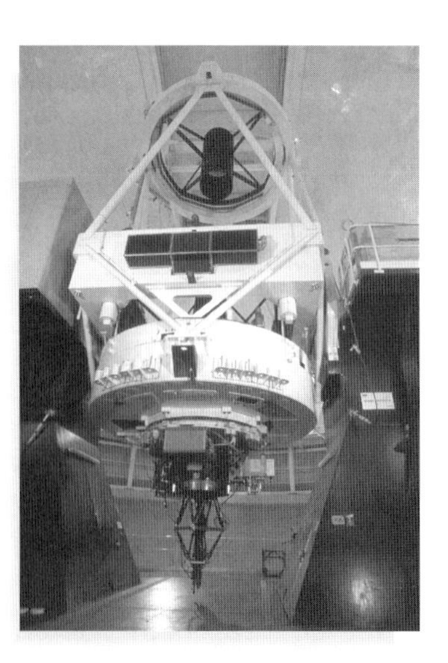

The William Herschel telescope. (Image reproduced by courtesy of Rainer Gernstein.)

Willstrop three-mirror camera

A wide angle (5°) camera design that uses three aspherical mirrors. The focal surface is curved and the outer parts of the image are *vignetted*, but unlike the *Schmidt camera*, it potentially can be made as large as any reflecting telescope.

window

1 A cover for the entrance aperture to an instrument that is transparent to the operating wavelengths.
2 See *atmospheric window*.

winter solstice

See *solstice*.

wire grid polarizer

A *polarizer* for use at infrared and longer wavelengths. It consists of a grid of electrically conducting parallel wires spaced at about five times their thickness. The device transmits linearly polarized radiation parallel to the wires.

Wise observatory

An optical observatory in Mitzpe Ramon. Israel.

WIYN observatory

An optical observatory with a 3.5m telescope on *Kitt Peak*.

W.M. Keck observatory

See *Keck telescopes*.

Wollaston prism

A *polarizer* that produces two diverging beams that are linearly polarized at right angles to each other. It consists of two right-angled prisms with orthogonal optical axes, cemented together along their long sides. See also: *beam splitter*.

Wolter telescope

See *grazing-incidence telescope*.

Wood's anomaly

See *ghost*.

work function

The minimum energy needed to cause the emission of electrons via the *photoelectric effect*.

Wright telescope

A variant of the *Schmidt camera* that employs an oblate spheroidal primary mirror. It has a flat focal surface.

Wyoming observatory

An infrared observatory sited at an altitude of 2900 m on Mount Jelm in Wyoming, with a 2.3m telescope.

X-ray

A short *wavelength*, high *frequency*, form of *electromagnetic radiation*, but less extreme than *gamma rays*. The wavelength ranges from 1 nm to 0.01 nm and the frequency from 3 x 10^{17} Hz to 3 x 10^{19} Hz. In energy terms, X-ray photons have energies between 10^3 eV and 10^5 eV. See also: *hard X-ray; soft-X-ray.*

X-ray calorimeter

A detector for X-rays that operates by absorbing the X-ray and detecting the resulting temperature rise. It is intrinsically sensitive to the X-ray's energy.

X-ray detector

A detector for *X-rays*. Since the Earth's atmosphere is opaque to X-ray radiation, X-ray detectors have to be flown on rockets or spacecraft. They can also normally be used to detect gamma rays and cosmic rays and include; *proportional counters, scintillation detectors, CCDs, solid-state detectors, nuclear emulsions, spark detectors* and *X-ray calorimeters.*

X-ray spectrometer

An instrument for discriminating between X-rays of differing energies. Many *X-ray detectors* are intrinsically sensitive to the X-ray's energy. Otherwise *diffraction gratings* may be used, either as *transmission gratings* or grazing-incidence *reflection gratings*. At higher energies, spectrometers depend upon Bragg reflection from crystals such as lithium fluoride or potassium acid phthalate (KAP). See also: *CCD.*

X-ray telescope

For low energy X-rays, relatively conventional reflecting telescopes may be used with multilayer coatings acting as the reflectors. These coatings are numerous alternate layers of, for example, tungsten and carbon, and provide X-ray reflectivities of up to 20%.

At higher energies, X-ray reflection only occurs for very shallow angles of incidence to the surface, and results in *grazing-incidence telescopes.*

XUV region

See *extreme ultraviolet region.*

Y

Yagi antenna

A radio *antenna* that comprises a *half-wave dipole* as the detecting element, with one or more rods placed behind and in front of it. The additional rods are made of a conducting material but are not electrically connected to the dipole (they are thus 'parasitic' leading to the alternative name 'parasitic aerial' for the device). The rods behind the dipole reflect radiation onto it and screen it from radio sources behind it. The rods in front of the dipole act to concentrate the radiation, and are called directors. The Yagi antenna is very familiar as the domestic TV aerial. In *radio telescopes* it may be used as the *feed* of a large radio dish, or many Yagis may be combined into an *array*.

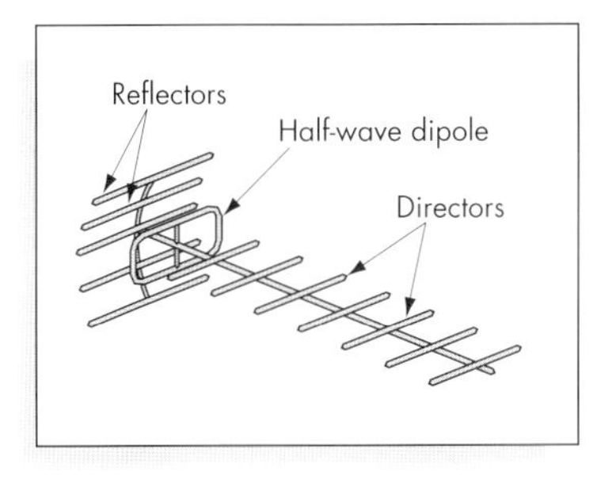

The Yagi antenna.

Yale bright star catalog

See *bright star catalog.*

Yearbook of Astronomy

A book published annually containing articles and listings of astronomical events for the year.

Yepun

See *very large telescope.*

Yerkes refractor

The largest refracting telescope in the world, with a diameter of 40 inches (1 m), completed in 1897. It was named for Charles Yerkes (1837–1905) who donated the money for its construction. It is sited at the Yerkes observatory at Williams Bay, Wisconsin, and operated by the University of Chicago. See also: *Lick refractor.*

The 40-inch (1 m) refractor at Yerkes.

Yerkes system

A synonym for the MK *spectral type.*

yoke mounting

See *English mounting.*

Yolo telescope

See *Schiefspiegler telescope.*

Z

Zeeman effect

The splitting of *spectrum lines* when the atoms emitting or absorbing them are in the presence of a magnetic field. The Zeeman components are *polarized* and this enables the strength and direction of the magnetic field producing them to be inferred. See also: *Stark effect; magnetometer.*

Zeiss

A commercial firm manufacturing telescopes and accessories and planetaria.

Zelenchukskaya observatory

A radio and optical observatory sited at Zelenchukskaya in the Caucasus at an altitude of 2100 m on Mount Pastukhov. It houses the 6m *Bolshoi telescop azimutalnyi* and the *RATAN 600* radio telescope.

zenith

The point directly overhead for an observer, opposite to the *nadir* in the sky.

zenithal hourly rate

The number of meteors in a meteor shower that would be seen from a good observing site if the shower's radiant was at the zenith. The observed rate will usually be lower than this figure.

zenith distance

The angular distance of an object in the sky from the *zenith*. It is equal to 90° – altitude.

zenith telescope

A vertically pointing fixed telescope used for accurate measurements of the positions of stars that pass close to the *zenith* and to calibrate clocks. See also: *photographic zenith tube.*

Zerodur

A low thermal expansion glass-ceramic material, widely used to form mirrors for large telescopes.

zero error

The reading given by an instrument for zero input. It can normally be measured and subtracted from the measured readings to give the correct result. See also: *dark exposure.*

zero image shift focuser

See *focuser.*

zero power finder

A *finding* system that does not magnify the image. At its simplest it may be like a "notch and bead" rifle sight. More complex devices, such as the '*Telrad*' finder, project an illuminated circle or dot via a beam splitter so that it appears against the sky for the observer to sight onto.

zodiac

See panel (pages 258–259).

zodiacal light

A faint elongated glow running along the *ecliptic* that results from solar light scattered by dust particles between the planets. It is best seen near the horizon just after sunset or just before dawn (when it is sometimes called the *false dawn*).

zone of avoidance

A band of sky between about 10° and 40° wide, centered on the Milky Way, within which *interstellar absorption* means that few galaxies may be seen.

zoom

A means of enlarging or diminishing the angular size of an image in a continuous fashion. Most *image processing* packages have a zoom facility. It is possibly to purchase zoom *eyepieces*, but their image quality is generally poorer than that of a fixed-focus eyepiece, and they tend to be very heavy.

Zwicky catalog

A catalog of some 40,000 galaxies and clusters of galaxies derived from the *Palomar observatory sky survey*, published in 1968 under the leadership of Fritz Zwicky. Its official tile is "Catalog of Galaxies and Clusters of Galaxies".

zodiac

A band of the sky about 20° wide, centered on the *ecliptic*, within which are found the Sun, Moon and planets. Traditionally there are twelve constellations to be found within the zodiac, but the modern constellation boundaries place a thirteenth, Ophiuchus, into the region as well. The full list of zodiacal constellations is thus:

Constellation	*Dates of solar passage*
Sagittarius	Dec 19 to Jan 21
Capricornus	Jan 22 to Feb 16
Aquarius	Feb 17 to Mar 12
Pisces	Mar 13 to Apr 18
Aries	Apr 19 to May 14
Taurus	May 15 to June 21
Gemini	June 22 to July 21
Cancer	July 22 to Aug 11
Leo	Aug 12 to Sept 17
Virgo	Sept 18 to Oct 31
Libra	Nov 1 to Nov 22
Scorpius	Nov 23 to Nov 30
Ophiuchus	Dec 1 to Dec 18.

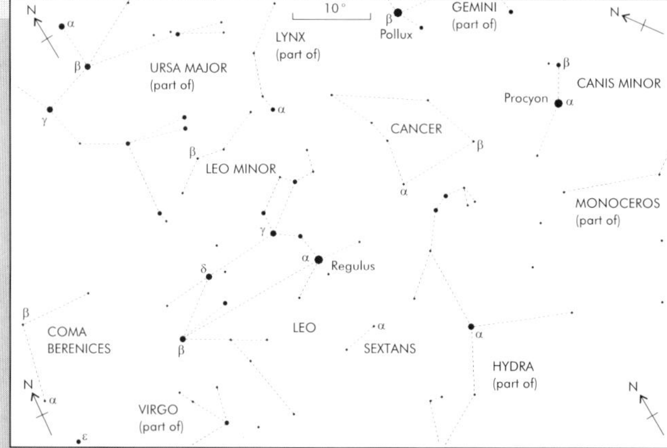

The Zodiacal constellations Leo and Cancer, plus parts of Virgo and Gemini.

Z

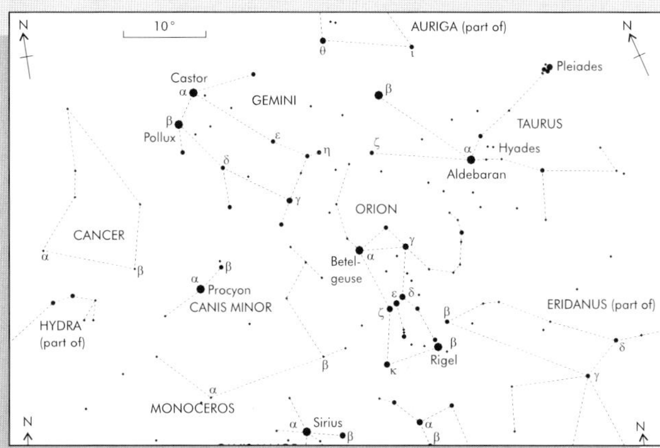

The Zodiacal constellations Cancer, .Gemini and Taurus

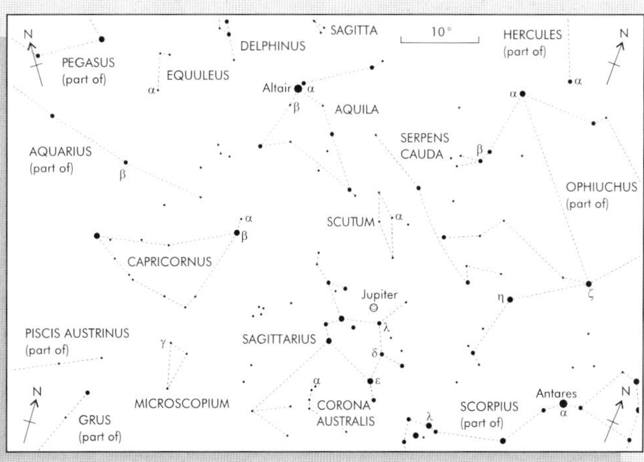

The Zodiacal constellations Capricornus and Sagittarius, plus parts of Aquarius, Ophiuchus and Scorpius.

Z

Appendix
Additional tables

Contents

A1 Messier catalogue

Object	Nebula name	NGC	Type	RA_{2000} H m	Dec_{2000} degrees	Constell-ation	Visual[1] mag.	Size[2] (')
Messier 1	Crab	1952	Supernova Remnant	05 35	22.0	Tau	8.4	6
Messier 2		7089	Globular Cluster	21 34	–0.8	Aqr	6.5	13
Messier 3		5272	Globular Cluster	13 42	28.4	CVn	6.4	16
Messier 4		6121	Globular Cluster	16 24	–26.5	Sco	5.9	26
Messier 5		5904	Globular Cluster	15 19	2.1	Ser Cap	5.8	17
Messier 6	Butterfly	6405	Open Cluster	17 40	–32.2	Sco	4.2	15
Messier 7		6475	Open Cluster	17 54	–34.8	Sco	3.3	80
Messier 8	Lagoon	6523	Emission Nebula	18 04	–24.4	Sgr	5.8	90
Messier 9		6333	Globular Cluster	17 19	–18.5	Oph	7.9	9
Messier 10		6254	Globular Cluster	16 57	–4.1	Oph	6.6	15
Messier 11		6705	Open Cluster	18 51	–6.3	Sct	5.8	14
Messier 12		6218	Globular Cluster	16 47	–2.0	Oph	6.6	15
Messier 13		6205	Globular Cluster	16 42	36.5	Her	5.9	17
Messier 14		6402	Globular Cluster	17 38	–3.3	Oph	7.6	12
Messier 15		7078	Globular Cluster	21 30	12.2	Peg	6.4	12
Messier 16	Eagle	6611	Open Cluster	18 19	–13.8	Ser	6.0	7
Messier 17	Omega	6618	Emission Nebula	18 21	–16.2	Sgr	6.0	46
Messier 18		6613	Open Cluster	18 20	–17.1	Sgr	6.9	9
Messier 19		6273	Globular Cluster	17 03	–26.3	Oph	7.2	14
Messier 20	Trifid	6514	Emission Nebula	18 03	–23.0	Sgr	6.3	29
Messier 21		6531	Open Cluster	18 05	–22.5	Sgr	5.9	13
Messier 22		6656	Globular Cluster	18 36	–23.9	Sgr	5.1	24
Messier 23		6494	Open Cluster	17 57	–19.0	Sgr	5.5	27
Messier 24		6603	Open Cluster	18 18	–18.4	Sgr	11.1	5
Messier 25		IC 4725	Open Cluster	18 32	–19.3	Sgr	4.6	32
Messier 26		6694	Open Cluster	18 45	–9.4	Sct	8.0	15
Messier 27	Dumbbell	6853	Planetary Nebula	20 00	22.7	Vul	7.6	15
Messier 28		6626	Globular Cluster	18 25	–24.9	Sgr	6.9	11
Messier 29		6913	Open Cluster	20 24	38.5	Cyg	6.6	7
Messier 30		7099	Globular Cluster	21 40	–23.2	Cap	7.5	11
Messier 31	Andromeda	224	Galaxy	00 43	41.3	And	3.5	180
Messier 32		221	Galaxy	00 43	40.9	And	8.2	8
Messier 33	Triangulum	598	Galaxy	01 34	30.7	Tri	5.7	62
Messier 34		1039	Open Cluster	02 42	42.8	Per	5.2	35
Messier 35		2168	Open Cluster	06 09	24.3	Gem	5.1	28
Messier 36		1960	Open Cluster	05 36	34.1	Aur	6.0	12
Messier 37		2099	Open Cluster	05 52	32.6	Aur	5.6	24
Messier 38		1912	Open Cluster	05 29	35.8	Aur	6.4	21
Messier 39		7092	Open Cluster	21 32	48.4	Cyg	4.6	32
Messier 40			2 stars	12 22	58.1	UMa	0.8	9
Messier 41		2287	Open Cluster	04 47	–20.7	CMa	4.5	38
Messier 42	Orion	1976	Emission Nebula	05 35	–5.5	Ori	4.0	66
Messier 43	Orion	1982	Emission Nebula	05 36	–5.3	Ori	9.0	20
Messier 44	Praesepe	2632	Open Cluster	08 40	20.0	Cnc	3.1	95
Messier 45	Pleiades		Open Cluster	03 47	24.1	Tau	1.2	110
Messier 46		2437	Open Cluster	07 42	–14.8	Pup	6.1	27

[1]This is the integrated magnitude over the whole area of the object. An angularly large object with a bright magnitude may therefore be less easy to see than a smaller object with a fainter magnitude. The magnitudes of the emission nebulae in particular may be misleading because they frequently contain brighter and darker regions.
[2]This is the largest dimension of the object. Some objects may be filamentary or have a brighter core or outer region making them easier to see than might be expected

A1 Messier catalogue (continued)

Object	Nebula name	NGC	Type	RA_{2000} H m	Dec_{2000} degrees	Constell-ation	Visual[1] mag.	Size[2] (')
Messier 47		2422	Open Cluster	07 37	–14.5	Pup	4.4	30
Messier 48		2548	Open Cluster	08 14	–5.8	Hya	5.8	54
Messier 49		4472	Galaxy	12 30	8.0	Vir	8.4	9
Messier 50		2323	Open Cluster	07 03	–8.3	Mon	5.9	16
Messier 51	Whirlpool	5194	Galaxy	13 30	47.2	CVn	8.4	11
Messier 52		7654	Open Cluster	23 24	61.6	Cas	6.9	13
Messier 53		5024	Globular Cluster	13 13	18.2	Com	7.7	13
Messier 54		6715	Globular Cluster	18 55	–30.5	Sgr	7.7	9
Messier 55		6809	Globular Cluster	19 40	–31.0	Sgr	7.0	19
Messier 56		6779	Globular Cluster	19 17	30.2	Lyr	8.3	7
Messier 57	Ring	6720	Planetary Nebula	18 54	33.0	Lyr	9.7	3
Messier 58		4579	Galaxy	12 38	11.8	Vir	9.8	5
Messier 59		4621	Galaxy	12 42	11.7	Vir	9.8	5
Messier 60		4649	Galaxy	12 44	11.6	Vir	8.8	7
Messier 61		4303	Galaxy	12 22	4.5	Vir	9.7	6
Messier 62		6266	Globular Cluster	17 01	–30.1	Oph	6.6	14
Messier 63	Sunflower	5055	Galaxy	13 16	42.0	CVn	8.6	12
Messier 64	Black–eye	4826	Galaxy	12 57	21.7	Com	8.5	9
Messier 65		3623	Galaxy	11 19	13.1	Leo	9.3	10
Messier 66		3627	Galaxy	11 20	13.0	Leo	9.0	9
Messier 67		2682	Open Cluster	08 50	11.8	Cnc	6.9	30
Messier 68		4590	Globular Cluster	12 40	–26.8	Hya	8.2	12
Messier 69		6637	Globular Cluster	18 31	–32.4	Sgr	7.7	7
Messier 70		6681	Globular Cluster	18 43	–32.3	Sgr	8.8	8
Messier 71		6838	Globular Cluster	19 54	18.8	Sge	8.3	7
Messier 72		6981	Globular Cluster	20 54	–12.5	Aqr	9.4	6
Messier 73		6994	Open Cluster	20 59	–12.6	Aqr	8.9	3
Messier 74		628	Galaxy	01 37	15.8	Psc	9.2	10
Messier 75		6864	Globular Cluster	20 06	–21.9	Sgr	8.6	6
Messier 76	Little Dumbbell	650	Planetary Nebula	01 42	51.6	Per	12.2	5
Messier 77		1068	Galaxy	02 43	0.0	Cet	8.8	7
Messier 78		2068	Reflection Nebula	05 47	0.1	Ori	8.0	8
Messier 79		1904	Globular Cluster	05 25	–24.6	Lep	8.0	9
Messier 80		6093	Globular Cluster	16 17	–23.0	Sco	7.2	9
Messier 81	Bode's	3031	Galaxy	09 56	69.1	UMa	6.9	26
Messier 82		3034	Galaxy	09 56	69.7	UMa	8.4	11
Messier 83		5236	Galaxy	13 37	–29.9	Hya	8.2	11
Messier 84		4374	Galaxy	12 25	12.9	Vir	9.3	5
Messier 85		4382	Galaxy	12 25	18.2	Com	9.2	7
Messier 86		4406	Galaxy	12 26	13.0	Vir	9.2	7
Messier 87	Virgo A	4486	Galaxy	12 31	12.4	Vir	8.6	7
Messier 88		4501	Galaxy	12 32	14.4	Com	9.5	7
Messier 89		4552	Galaxy	12 36	12.6	Vir	9.8	4
Messier 90		4569	Galaxy	12 37	13.2	Vir	9.5	10
Messier 91		4548	Galaxy	12 35	14.5	Com	10.2	5
Messier 92		6341	Globular Cluster	17 17	43.1	Her	6.5	11
Messier 93		2447	Open Cluster	07 45	–23.9	Pup	6.2	22
Messier 94		4736	Galaxy	12 51	41.1	CVn	8.2	11
Messier 95		3351	Galaxy	10 44	11.7	Leo	9.7	7

A1 Messier catalogue (continued)

Object	Nebula name	NGC	Type	RA_{2000} H m	Dec_{2000} degrees	Constell-ation	Visual[1] mag.	Size[2] (')
Messier 96		3368	Galaxy	10 47	11.8	Leo	9.2	7
Messier 97	Owl	3587	Planetary Nebula	11 15	55.0	UMa	12	3
Messier 98		4192	Galaxy	12 14	14.9	Com	10.1	10
Messier 99		4254	Galaxy	12 19	14.4	Com	9.8	5
Messier 100		4321	Galaxy	12 23	15.8	Com	9.4	7
Messier 101	Pinwheel	5457	Galaxy	14 03	54.4	UMa	7.7	27
Messier 102		5866	Galaxy	15 07	55.8	Dra	10.0	5
Messier 103		581	Open Cluster	01 33	60.7	Cas	7.4	6
Messier 104	Sombrero	4594	Galaxy	12 40	−11.6	Vir	8.3	9
Messier 105		3379	Galaxy	10 48	12.6	Leo	9.3	5
Messier 106		4258	Galaxy	12 19	47.3	CVn	8.3	18
Messier 107		6171	Globular Cluster	16 33	−13.1	Oph	8.1	10
Messier 108		3556	Galaxy	11 12	55.7	UMa	10.1	8
Messier 109		3992	Galaxy	11 58	53.4	UMa	9.8	8

(Data for this table obtained from Sky Catalogue 2000.0 Vol 2 (Ed. A. Hirshfeld & R.W. Sinnott, Cambridge University Press 1985), Astrophysical Quantities (C.W. Allen, Athlone Press 1973, NGC 2000.0, (R.W. Sinnott, Cambridge University Press, 1988), Visual Astronomy of the Deep Sky, (R.N. Clark, Cambridge University Press, 1990), Hartung's Astronomical Objects for Southern Telescopes, (D. Malin, D.J. Frew, Cambridge University Press, 1995), Astrophysical Journal Supplement, **4**, 257, 1959, S. Sharpless.)

A2 Caldwell catalogue

Object	Nebula name	NGC	Type	RA_{2000} H m	Dec_{2000} degrees	Constell-ation	Visual[3] mag.	Size[4] (')
Caldwell 1		188	Open Cluster	00 44	85.3	Cep	8.1	14
Caldwell 2		40	Planetary Nebula	00 13	72.5	Cep	10.7	0.6
Caldwell 3		4236	Galaxy	12 17	69.5	Dra	9.7	19
Caldwell 4		7023	Reflection Nebula	21 02	68.2	Cep	7.0	18
Caldwell 5		IC 342	Galaxy	03 47	68.1	Cam	9.1	18
Caldwell 6	Cat's Eye	6543	Planetary Nebula	17 59	66.6	Dra	8.8	6
Caldwell 7		2403	Galaxy	07 37	65.6	Cam	8.4	18
Caldwell 8		559	Open Cluster	01 30	63.3	Cas	9.5	5
Caldwell 9	Cave	Sh2-155	Emission Nebula	22 57	62.6	Cep	≈ 9	50
Caldwell 10		663	Open Cluster	01 46	61.3	Cas	7.1	16
Caldwell 11	Bubble	7635	Emission Nebula	23 21	61.2	Cas	8.5	15
Caldwell 12		6946	Galaxy	20 35	60.2	Cep	8.9	11
Caldwell 13		457	Open Cluster	01 19	58.3	Cas	6.4	13
Caldwell 14	h & chi Per	869 / 884	Open Cluster	02 20	57.1	Per	4.3/4.4	30/30
Caldwell 15	Blinking	6826	Planetary Nebula	19 45	50.5	Cyg	9.8	2
Caldwell 16		7243	Open Cluster	22 15	49.9	Lac	6.4	21
Caldwell 17		147	Galaxy	00 33	48.5	Cas	9.3	13
Caldwell 18		185	Galaxy	00 39	48.3	Cas	9.2	12
Caldwell 19	Cocoon	IC 5146	Emission Nebula	21 54	47.3	Cyg	7.2	12
Caldwell 20	North America	7000	Emission Nebula	20 59	44.3	Cyg	5.0	120
Caldwell 21		4449	Galaxy	12 28	44.1	CVn	9.4	5
Caldwell 22	Blue Snowball	7662	Planetary Nebula	23 26	42.6	And	9.2	2
Caldwell 23		891	Galaxy	02 23	42.4	And	10.0	14
Caldwell 24		1275	Galaxy	03 20	41.5	Per	11.6	3
Caldwell 25		2419	Globular Cluster	07 38	38.9	Lyn	10.4	4
Caldwell 26		4244	Galaxy	12 18	37.8	CVn	10.2	16
Caldwell 27	Crescent	6888	Emission Nebula	20 12	38.4	Cyg	≈ 11	20
Caldwell 28		752	Open Cluster	01 58	37.7	And	5.7	50
Caldwell 29		5005	Galaxy	13 11	37.1	CVn	9.8	5
Caldwell 30		7331	Galaxy	22 37	34.4	Peg	9.5	11
Caldwell 31	Flaming Star	IC 405	Emission Nebula	05 16	34.3	Aur	≈ 7	30
Caldwell 32		4631	Galaxy	12 42	32.5	CVn	9.3	15
Caldwell 33	Veil (E)	6992 / 5	Supernova Remnant	20 57	31.5	Cyg	8.0	60
Caldwell 34	Veil (W)	6960	Supernova Remnant	20 46	30.7	Cyg	8.0	70
Caldwell 35		4889	Galaxy	13 00	28.0	Com	11.4	3
Caldwell 36		4559	Galaxy	12 36	28.0	Com	9.9	11
Caldwell 37		6885	Open Cluster	20 12	26.5	Vul	5.7	7
Caldwell 38		4565	Galaxy	12 36	26.0	Com	9.6	16
Caldwell 39	Eskimo	2392	Planetary Nebula	07 29	20.9	Gem	9.9	0.7
Caldwell 40		3626	Galaxy	11 20	18.4	Leo	10.9	3
Caldwell 41	Hyades		Open Cluster	04 27	16.0	Tau	0.5	330
Caldwell 42		7006	Globular Cluster	21 02	16.2	Del	10.6	3
Caldwell 43		7814	Galaxy	00 03	16.2	Peg	10.5	6
Caldwell 44		7479	Galaxy	23 05	12.3	Peg	11.0	4
Caldwell 45		5248	Galaxy	13 38	8.9	Boπ	10.2	7

[3]This is the integrated magnitude over the whole area of the. An angularly large object with a bright magnitude may therefore be less easy to see than a smaller object with a fainter magnitude. The magnitudes of the emission nebulae in particular may be misleading because they frequently contain brighter and darker regions. The symbol "≈" indicates a magnitude estimated from visual descriptions.
[4]This is the largest dimension of the object. Some objects may be filamentary or have a brighter core or outer region making them easier to see than might be expected

A2 Caldwell catalogue (continued)

Object	Nebula name	NGC	Type	RA_{2000} H m	Dec_{2000} degrees	Constell-ation	Visual[3] mag.	Size[4] (')
Caldwell 46	Hubble's variable	2261	Emission Nebula	06 39	8.7	Mon	10.0	2
Caldwell 47		6934	Globular Cluster	20 34	7.4	Del	8.9	6
Caldwell 48		2775	Galaxy	09 10	7.0	Cnc	10.3	5
Caldwell 49	Rosette	2237–9	Emission Nebula	06 32	5.1	Mon	≈ 4	80
Caldwell 50		2244	Open Cluster	06 32	4.9	Mon	4.8	24
Caldwell 51		IC 1613	Galaxy	01 05	2.1	Cet	9.3	12
Caldwell 52		4697	Galaxy	12 49	–5.8	Vir	9.3	6
Caldwell 53	Spindle	3115	Galaxy	10 05	–7.7	Sex	9.2	8
Caldwell 54		2506	Open Cluster	08 00	–10.8	Mon	7.6	7
Caldwell 55	Saturn	7009	Planetary Nebula	21 04	–11.4	Aqr	8.3	2
Caldwell 56		246	Planetary Nebula	00 47	–11.9	Cet	8.0	4
Caldwell 57	Barnard's	6822	Galaxy	19 45	–14.8	Sgr	9.4	10
Caldwell 58		2360	Open Cluster	07 18	–15.6	CMa	7.2	13
Caldwell 59	Ghost of Jupiter	3242	Planetary Nebula	10 25	–18.6	Hya	8.6	21
Caldwell 60	Antennae	4038	Galaxy	12 02	–18.9	Crv	10.7	3
Caldwell 61	Antennae	4039	Galaxy	12 02	–18.9	Crv	10.7	3
Caldwell 62		247	Galaxy	00 47	–20.8	Cet	8.9	20
Caldwell 63	Helix	7293	Planetary Nebula	22 30	–20.8	Aqr	7.4	13
Caldwell 64		2362	Open Cluster	07 19	–25.0	CMa	4.1	8
Caldwell 65	Silver Coin	253	Galaxy	00 48	–25.3	Scl	7.1	25
Caldwell 66		5694	Globular Cluster	14 40	–26.5	Hya	10.2	4
Caldwell 67		1097	Galaxy	02 46	–30.3	For	9.3	9
Caldwell 68	R CrA	6729	Reflection Nebula	19 02	–37.0	CrA	≈ 11	1
Caldwell 69	Bug	6302	Planetary Nebula	17 14	–37.1	Sco	12.8	1
Caldwell 70		300	Galaxy	00 55	–37.7	Scl	8.7	20
Caldwell 71		2477	Open Cluster	07 52	–38.6	Pup	5.8	27
Caldwell 72		55	Galaxy	00 15	–39.2	Scl	7.9	32
Caldwell 73		1851	Globular Cluster	05 14	–40.1	Col	7.3	11
Caldwell 74	Eight–Burst	3132	Planetary Nebula	10 08	–40.4	Vel	8.2	0.8
Caldwell 75		6124	Open Cluster	16 26	–40.7	Sco	5.8	29
Caldwell 76		6231	Open Cluster	16 54	–41.8	Sco	2.6	15
Caldwell 77	Cen A	5128	Galaxy	13 26	–43.0	Cen	7.0	18
Caldwell 78		6541	Globular Cluster	18 08	–43.7	CrA	6.6	13
Caldwell 79		3201	Globular Cluster	10 18	–46.4	Vel	6.8	18
Caldwell 80	Omega Centauri	5139	Globular Cluster	13 27	–47.5	Cen	3.7	36
Caldwell 81		6352	Globular Cluster	17 26	–48.4	Ara	8.2	7
Caldwell 82		6193	Open Cluster	16 41	–48.8	Ara	5.2	15
Caldwell 83		4945	Galaxy	13 05	–49.5	Cen	8.6	20
Caldwell 84		5286	Globular Cluster	13 46	–51.4	Cen	7.6	9
Caldwell 85		IC 2391	Open Cluster	08 40	–53.1	Vel	2.5	50
Caldwell 86		6397	Globular Cluster	17 41	–53.7	Ara	5.7	26
Caldwell 87		1261	Globular Cluster	03 12	–55.2	Hor	8.4	7
Caldwell 88		5823	Open Cluster	15 06	–53.6	Cir	7.9	10
Caldwell 89		6067	Open Cluster	16 13	–54.2	Nor	5.6	13
Caldwell 90		2867	Planetary Nebula	09 21	–58.3	Car	9.7	0.2
Caldwell 91		3532	Open Cluster	11 06	–58.7	Car	3.0	55
Caldwell 92	Eta Carina	3372	Emission Nebula	10 44	–59.9	Car	2.5	120
Caldwell 93		6752	Globular Cluster	19 11	–60.0	Pav	5.4	20
Caldwell 94	Jewel Box	4755	Open Cluster	12 54	–60.3	Cru	4.2	10
Caldwell 95		6025	Open Cluster	16 04	–60.5	TrA	5.1	12

A2 Caldwell catalogue (continued)

Object	Nebula name	NGC	Type	RA_{2000}	Dec_{2000}	Constell–	Visual[3]	Size[4]
Caldwell 96		2516	Open Cluster	07 58	-60.9	Car	3.8	30
Caldwell 97		3766	Open Cluster	11 36	-61.6	Cen	5.3	12
Caldwell 98		4609	Open Cluster	12 42	-63.0	Cru	6.9	5
Caldwell 99	Coalsack		Absorption Nebula	12 53	-63.0	Cru	—	350
Caldwell 100		IC 2944	Open Cluster	11 37	-63.0	Cen	4.5	15
Caldwell 101		6744	Galaxy	19 10	-63.9	Pav	8.4	16
Caldwell 102	Southern Pleiades	IC 2602	Open Cluster	10 43	-64.4	Car	1.9	50
Caldwell 103	Tarantula	2070	Emission Nebula	05 39	-69.1	Dor	8.2	40
Caldwell 104		362	Globular Cluster	01 03	-70.9	Tuc	6.6	13
Caldwell 105		4833	Globular Cluster	13 00	-70.9	Mus	7.4	14
Caldwell 106	47 Tucanae	104	Globular Cluster	00 24	-72.1	Tuc	4.0	31
Caldwell 107		6101	Globular Cluster	16 26	-72.2	Aps	9.3	11
Caldwell 108		4372	Globular Cluster	12 26	-72.7	Mus	7.8	19
Caldwell 109		3195	Planetary Nebula	10 10	-80.9	Cha	11.6	0.6

(Data for this table obtained from Sky Catalogue 2000.0 Vol 2 (Ed. A. Hirshfeld & R.W. Sinnott, Cambridge University Press 1985), the Caldwell Card (Sky Publishing Corp. 1996), NGC 2000.0, (R.W. Sinnott, Cambridge University Press, 1988), Visual Astronomy of the Deep Sky, (R.N. Clark, Cambridge University Press, 1990), Hartung's Astronomical Objects for Southern Telescopes, (D. Malin, D.J. Frew, Cambridge University Press, 1995), Astrophysical Journal Supplement, **4**, 257, 1959, S. Sharpless.)

A3 Constellations

Constellation	Abbreviation
Andromeda	And
Antlia	Ant
Apus	Aps
Aquarius	Aqr
Aquila	Aql
Ara	Ara
Aries	Ari
Auriga	Aur
Boπtes	Boo
Caelum	Cae
Camelopardalis	Cam
Cancer	Cnc
Canes Venatici	CVn
Canis Major	CMa
Canis Minor	CMi
Capricornus	Cap
Carina	Car
Cassiopeia	Cas
Centaurus	Cen
Cepheus	Cep
Cetus	Cet
Chamaeleon	Cha
Circinus	Cir
Columba	Col
Coma Berenices	Com
Corona Australis	CrA
Corona Borealis	CrB
Corvus	Crv
Crater	Crt
Crux	Cru
Cygnus	Cyg
Delphinus	Del
Dorado	Dor
Draco	Dra
Equuleus	Equ
Eridanus	Eri
Fornax	For
Gemini	Gem
Grus	Gru
Hercules	Her
Horologium	Hor
Hydra	Hya
Hydrus	Hyi
Indus	Ind
Lacerta	Lac
Leo	Leo
Leo Minor	LMi
Lepus	Lep
Libra	Lib
Lupus	Lup
Lynx	Lyn
Lyra	Lyr
Mensa	Men
Microscopium	Mic
Monoceros	Mon
Musca	Mus
Norma	Nor
Octans	Oct
Ophiuchus	Oph
Orion	Ori
Pavo	Pav
Pegasus	Peg
Perseus	Per
Phoenix	Phe
Pictor	Pic
Pisces	Psc
Piscis Austrinus	PsA
Puppis	Pup
Pyxis	Pyx
Reticulum	Ret
Sagitta	Sge
Sagittarius	Sgr
Scorpius	Sco
Sculptor	Scl
Scutum	Sct
Serpens	Ser
Sextans	Sex
Taurus	Tau
Telescopium	Tel
Triangulum	Tri
Triangulum Australe	TrA
Tucana	Tuc
Ursa Major	UMa
Ursa Minor	UMi
Vela	Vel
Virgo	Vir
Volans	Vol
Vulpecula	Vul

A4 Julian date

Date (Midday, Jan 1st, Gregorian calendar)	Julian date
2050	2469808.0
2025	2460677.0
2000	2451545.0
1975	2442414.0
1950	2433283.0
1925	2424152.0
1900	2415021.0
1850	2396759.0
1800	2378497.0
1750	2360234.0
1700	2341972.0
1650	2323710.0
1600	2305448.0
1200	2159351.0
800	2013254.0
400	1867157:0
0	1721060.0
400BC	1574963.0
800 BC	1428866.0
1200 BC	1282769.0
1600 BC	1136672.0
2000 BC	990575.0
4000 BC	260090.0
4714 BC (24 Nov)	0.0

Running number of days throughout a year

Date (Midday)	Normal year	Leap year
1 Feb	31	31
1 Mar	59	60
1 Apr	90	91
1 May	120	121
1 Jun	151	152
1 Jul	181	182
1 Aug	212	213
1 Sep	243	244
1 Oct	273	274
1 Nov	304	305
1 Dec	334	335
1 Jan	365	366

A5 Radio astronomy allocated wavelength zones (WARC 1979)

37.5–38.25 MHz
73–74.6 MHz
150.05–153 MHz
322–328.6 MHz
406.1–410 MHz

608–614 MHz
1400–1427 MHz
1660–1670 MHz
2655–2700 MHz
4800–5000 MHz

10.6–10.7 GHz
14.47–14.5 GHz
15.35–15.4 GHz
22.21–22.5 GHz
23.6–24 GHz

31.1–31.8 GHz
42.5–43.5 GHz
86–92 GHz
105–116 GHz
182–185 GHz

217–231 GHz
265–267 GHz

A6 Greek alphabet

Letter	Lower case	Upper case
Alpha	α	Α
Beta	β	Β
Gamma	γ	Γ
Delta	δ	Δ
Epsilon	ε	Ε
Zeta	ζ	Ζ
Eta	η	Η
Theta	θ	Θ
Iota	ι	Ι
Kappa	κ	Κ
Lambda	λ	Λ
Mu	μ	Μ
Nu	ν	Ν
Xi	ξ	Ξ
Omicron	ο	Ο
Pi	π	Π
Rho	ρ	Ρ
Sigma	σ	Σ
Tau	τ	Τ
Upsilon	υ	Υ
Phi	φ	Φ
Chi	χ	Χ
Psi	ψ	Ψ
Omega	ω	Ω

A7 SI prefixes

Prefix	Multiplier	Symbol
atto	10^{-18}	a
femto	10^{-15}	f
pico	10^{-12}	p
nano	10^{-9}	η
micro	10^{-6}	μ
milli	10^{-3}	m
centi	10^{-2}	c (not recommended)
deci	10^{-1}	d (not recommended)
deca	10^{1}	da (not recommended)
hecto	10^{2}	h (not recommended)
kilo	10^{3}	k
mega	10^{6}	M
giga	10^{9}	G
tera	10^{12}	T
peta	10^{15}	P
exa	10^{18}	E

A8 SI Units

Physical quantity	Unit	Symbol
angle	radian	rad
capacitance	farad	F (s^4 A^2 m^{-2} kg^{-1})
electric charge	coulomb	C (A s)
electric current	ampere	A
electrical resistance	ohm	Ω (m^2 kg s^{-3} A^{-2})
energy	joule	J (m^2 kg s^{-2})
force	newton	N (kg m s^{-1})
frequency	hertz	Hz (s^{-1})
length	metre	m
luminous intensity	candela	cd
magnetic flux density	tesla	T (kg s^{-2} A^{-1})
mass	kilogram	kg
power	watt	W (m^2 kg s^{-3})
pressure	pascal	Pa (kg m^{-1} s^{-2})
solid angle	steradian	sr
temperature	kelvin	K
time	second	s
voltage	volt	V (m^2 kg s^{-3} A^{-1})

A9 Other units in common use in astronomy

Unit	Symbol	Equivalent
Ångstrom	Å	10^{-10} m
astronomical unit	AU	$1.49597870 \times 10^{11}$ m
atmosphere	atm	1.01325×10^{5} Pa
bar	bar	10^{5} Pa
dyne	dyn	10^{-5} N
electron volt	eV	1.6022×10^{-19} J
erg	erg	10^{-7} J
gauss	G	10^{-4} T
jansky	jy	10^{-26} W m^{-2} Hz^{-1}
light year	ly	9.4605×10^{15} m
micron	μ, μm	10^{-6} m
parsec	pc	3.0857×10^{16} m
solar luminosity	L_o	3.8478×10^{26} W
solar mass	M_o	1.9891×10^{30} kg
solar radius	R_o	6.960×10^{8} m

A10 Physical and Astronomical constants and symbols

amu	Atomic mass unit: one twelfth of the mass of the carbon-12 nucleus = 1.66053×10^{-27} kg
B	B magnitude in the UBV system; magnetic field strength
c	Velocity of light in vacuo = 2.9979250×10^{8} m s^{-1}
C_p C_v	Specific heats of a gas at constant pressure and constant volume
e^- e^+	Symbols for electrons and positrons
G	Gravitational constant = 6.670×10^{-11} N m^2 kg^{-2}
h	Planck's constant = 6.62620×10^{-34} J s
M	Absolute magnitude
m	Apparent magnitude
m_e	Mass of the electron = 9.10956×10^{-31} kg
m_p	Mass of the proton = 1.67352×10^{-27} kg
p^+	Symbol for a proton
U	U magnitude of the UBV system
V	V magnitude of the UB V system
X	Mass fraction of hydrogen
Y	Mass fraction of helium
Z	Mass fraction of all elements other than hydrogen and helium
α	Right ascension
δ	Declination
ε_o	Permittivity of a vacuum = 8.85×10^{-12} F m^{-1}
λ	Wavelength
μ	Molecular weight; refractive index; cos θ
ν	Frequency
ρ	Density
σ	Stefan's constant (Stefan–Boltzmann constant) = 5.6696 × 10–8 Wm^{-2} K^{-4} ; Standard deviation
τ	Optical depth

A11 Astronomy societies

For details of local astronomical societies see "International Directory of Astronomical Associations and Societies", published annually by the Centre de Donnés de Strasbourg, Université de Strasbourg.

Agrupación Astronáutica Española
Rosellón 134
E-08036 Barcelona
España

American Association for the Advancement of Science
1333 II Street NW
Washington DC 2005
USA

American Association of Variable Star Observers
25 Birch St.
Cambridge
MA 02138
USA

American Astronomical Society
2000 Florida Avenue NW
Suite 3000
Washington DC 20009
USA

Association Française d' Astronomie
Observatoire de Montsouris
17 Rue Emile-Deutsch-de-la-Meurthe
F-75014 Paris
France

Association des Groupes d' Astronomes Amateurs
4545 Avenue Pierre-de-Coubertin
Casier Postal 1000, Succ M.
Montreal
QC H1V 3R2
Canada

Association Nationale Science Techniques Jeunesse, Section Astronomique
Palais de la Découverte
Avenue, Franklin Roosevelt
F-75008 Paris
France

Association of Lunar and Planetary Observers
PO Box 16131
San Francisco
CA 94116
USA

Astronomical-Geodetical Society of Russia
24 Sadovaja-Kudrinskaya Ul.
SU-103101 Moskwa
Russia

Astronomical League
6235 Omie Circle
Pensacola
FL 32504
USA

Astronomical Society of Australia
School of Physics
University of Sydney
Sydney
NSW 2006
Australia

Astronomical Society of the Pacific
1290 24th Avenue
San Francisco
CA 94122
USA

Astronomical Society of Southern Africa
Southern African Astronomical Observatory
PO Box 9
Observatory 7935
South Africa

Astronomisk Selskab
Observatoriet
Øster Volgade 3
DK-1350 København K
Danmark

British Astronomical Association
Burlington House
Piccadilly
London W1V 9AG
United Kingdom

British Interplanetary Society
27/29 South Lambeth Rd.
London SW8 1SZ
United Kingdom

Canadian Astronomical Society
Dominion Astrophysical Observatory
5071 W. Saanich Rd.
Victoria
BC V8X 4M6
Canada

Committee on Space Research (COSPAR)
51 Bd de Montmorency
F-75016 Paris
France

Earthwatch
680 Mount Auburn St.
Box 403
Watertown
MA 02272
USA

Federation of Astronomical Societies
1 Tal-y-Bont Rd.
Ely
Cardiff, CF5 5EU
Wales

International Astronomical Union
61, Avenue de l' Observatoire
F-75014 Paris
France

Junior Astronomical Society
10 Swanwick Walk
Tadley
Basingstoke
Hampshire RG26 6JZ
United Kingdom

National Space Society
West Wing Suite 203
600 Maryland Avenue SW
Washington DC 20024
USA

Nederlandse Astronomenclub
Netherlands Foundation for Radio Astronomy
Postbus2
NL-7990 AA Dwingeloo
Nederland

Nederlandse Vereniging voor Weer-en Sterrenkunde
Nachtegaalstrat 82 bis
NL-3581 AN Utrecht
Nederland

Nippon Temmon Gakkai
Tokyo Tenmondai
2-21-1 Mitaka
Tokyo 181
Japan

Royal Astronomical Society
Burlington House
Piccadilly
London W1V 0NL
United Kingdom

Royal Astronomical Society of Canada
136 Dupont St.
Toronto
Ontario M5R 1V2
Canada

Royal Astronomical Society of New Zealand
PO Box 3181
Wellington
New Zealand

Schweizerische Astronomische Gesellschaft
Hirtenhoffstrasse 9
CH-6005 Luzern
Switzerland

Società Astronomica Italiana
Osservatorio Astrofisico di Arcetri
Largo E. Fermi 5
I-50125 Firenze
Italia

Société d'Astronomie Populaire
1 Avenue Camille Flammarion
F-31500 Toulouse
France

Société Astronomique de France
3 Rue Beethoven
F-75016 Paris
France

Société Royale Belge d' Astronomie, de Météorologie, et de Physique du Globe
Observatoire Royale de Belgique
Avenue Circulaire 3
B-1180 Bruxelles
Belgique

Stichting De Koepel
Nachtegaalstrat 82 bis
NL-3581 AN Utrecht
Nederland

Svenska Astronomiska Sallskapet
Stockholms Observatorium
S133 00 Saltsjöbaden
Sweden

Vercinigung der Sternfreunde e.V.
Volkssternwarte
Anzingerstrasse 1
D-8000 Munchen
Deutschland

Zentral Kommission Astronomie und Raumfahrt
Postfach 34
DDR-1030 Berlin
Deutschland

A12 Bibliography

Journals

Only the major and relatively widely available journals are listed. There are numerous more specialised research-level journals available in academic libraries, some of which have entries in the main part of the text.

Astronomy
Astronomy Now
Ciel et Espace
New Scientist
Scientific American
Sky and Telescope

Ephemerises

Astronomical Almanac (Published for each year), H.M.S.O. / U.S. Government Printing Office
Handbook of the British Astronomical Association (Published for each year), British Astronomical Association
Yearbook of Astronomy (Published each year) Macmillan

Star and other Catalogues, Atlases, Sky guides

Deep Sky Observer's Year, G. Privett, P. Parsons, Springer-Verlag, 2001
Deep Sky Observing, S.R. Coe, Springer-Verlag, 2000
Field Guide to the Deep Sky Objects, M. Inglis, Springer-Verlag, 2001
Messier's Nebulae and Star Clusters, K.G. Jones, Cambridge University Press, 1991
Norton's 2000.0, 19th Edition, I. Ridpath (Ed.), Longman, 1998
Observer's Sky Atlas, E. Karkoschka, Springer-Verlag, 1999
Observer's Year, P. Moore, Springer-Verlag, 1998
Observing Handbook and Catalogue of Deep Sky Objects, C. Luginbuhl, B. Skiff, Cambridge University Press, 1990
Observing the Caldwell Objects, D. Ratledge, Springer-Verlag, 2000
Photographic Atlas of the Stars. Arnold H, Doherty P, Moore P. IoP Publishing, 1997.
Photo-Guide to the Constellations, C.R. Kitchin, Springer-Verlag, 1997
Seeing Stars, C.R. Kitchin, R. Forrest, Springer-Verlag, 1997
Sky Atlas 2000.0, W. Tirion, Sky Publishing Corporation, 2000
Sky Catalogue 2000, Volumes 1 and 2, A. Hirshfield, R. W. Sinnott. Cambridge University Press, 1992

Reference Books

Cambridge Encyclopedia of the Sun, K.R. Lang, Cambridge University Press, 2001
Encyclopedia of Astronomy and Astrophysics, Edited by P. Murdin, Nature and IoP Publishing, 2001
Encyclopedia of Planetary Sciences, Edited by J.H. Shirley and R.W. Fairbridge, Kluwer Academic Publishers, 2000

Introductory Astronomy Books

AstroFAQs, S.F. Tonkin, Springer-Verlag, 2000
Astronomy; A self-teaching guide, D.L. Moche, John Wiley & Sons, 1993
Astronomy on the Personal Computer, O. Montenbruck, T. Pfleger, Springer-Verlag, 2000
Astronomy; The Evolving Universe, M. Zeilik, John Wiley, 1994
Astronomy through Space and Time, S. Engelbrektson, WCB, 1994
Eyes on the Universe, P. Moore, Springer-Verlag, 1997
Introductory Astronomy, K. Halliday, John Wiley, 1999

Introductory Astronomy and Astrophysics, M. Zeilik, S.A. Gregory, E.v.P. Smith, Saunders, 1992
Unfolding our Universe. I. Nicolson, Cambridge University Press, 1999
Universe. R.A.Freedman and W.J. Kaufmann III, W.H. Freeman, 2001

Practical Astronomy Books

Amateur Telescope Making, S.F. Tonkin, Springer-Verlag, 1999
Art and Science of CCD Astronomy, D. Ratledge, Springer-Verlag, 1997
Astronomical Equipment for Amateurs, M. Mobberley, Springer-Verlag, 1999
Astronomy with Small Telescopes, S.F. Tonkin, Springer-Verlag, 2001
Astrophysical Techniques, C.R. Kitchin, IoP Publishing, 1998
Building and Using an Astronomical Observatory. P. Doherty, Stevens, 1986
Challenges of Astronomy, W. Schlosser, T. Schmidt-Kaler, E.F. Malone, Springer Verlag, 1991.
Choosing and Using a Schmidt-Cassegrain Tel
2001
Compendium of Practical Astronomy. G.D. Roth,
Modern Amateur Astronomer P. Moore (Ed.) Spr
Observational Astronomy, D. S. Birney, Cambridg
Observing Meteors, Comets, Supernovae, N. Bon
Observing the Moon, P. Wlasuk, Springer-Verlag,
Practical Astronomy; A User Friendly Handbook
1993
Seeing the Sky; 100 Projects, Activities and Explor
Wiley & Sons, 1990.
Software and Data for Practical Astronomers, D. I
Solar Observing Techniques, C.R. Kitchin, Spring
Star Hopping; Your Visa to the Universe, R.A. Gar
1993
Telescope & Techniques, C. Kitchin, Springer 1995
Using the Meade ETX, M. Weasner, Springer-Verl